Springer Desktop Editions in Chemistry

L. Brandsma, S. F. Vasilevsky, H. D. Verkruijsse
Application of Transition Metal Catalysts in Organic Synthesis
ISBN 3-540-65550-6

H. Driguez, J. Thiem (Eds.)
Glycoscience, Synthesis of Oligosaccharides and Glycoconjugates
ISBN 3-540-65557-3

H. Driguez, J. Thiem (Eds.)
Glycoscience, Synthesis of Substrate Analogs and Mimetics
ISBN 3-540-65546-8

H. A. O. Hill, P. J. Sadler, A. J. Thomson (Eds.)
Metal Sites in Proteins and Models, Iron Centres
ISBN 3-540-65552-2

H. A. O. Hill, P. J. Sadler, A. J. Thomson (Eds.)
Metal Sites in Proteins and Models, Phosphatases, Lewis Acids and Vanadium
ISBN 3-540-65553-0

H. A. O. Hill, P. J. Sadler, A. J. Thomson (Eds.)
Metal Sites in Proteins and Models, Redox Centres
ISBN 3-540-65556-5

A. Manz, H. Becker (Eds.)
Microsystem Technology in Chemistry and Life Sciences
ISBN 3-540-65555-7

P. Metz (Ed.)
Stereoselective Heterocyclic Synthesis
ISBN 3-540-65554-9

H. Pasch, B. Trathnigg
HPLC of Polymers
ISBN 3-540-65551-4

T. Scheper (Ed.)
New Enzymes for Organic Synthesis, Screening, Supply and Engineering
ISBN 3-540-65549-2

D1640100

Springer
Berlin
Heidelberg
New York
Barcelona
Hong Kong
London
Milan
Paris
Singapore
Tokyo

H.A.O. Hill, P.J. Sadler, A.J. Thomson (Eds.)

Metal Sites in Proteins and Models
Phosphatases, Lewis Acids and Vanadium

 Springer

Prof. H. A. O. Hill
University of Oxford
Inorganic Chemistry Laboratory
South Park Road
OX1 3WR Oxford, Great Britain

Prof. P. J. Sadler
University of Edinburgh
Department of Chemistry
West Mains Road
EH9 3JJ Edinburgh, Great Britain
E-mail: P.J.Sadler@ed.ac.uk

Prof. A. J. Thomson
University of East Anglia
School of Chemical Sciences
NR4 7TJ Norwich, Great Britain

Description of the Series

The Springer Desktop Editions in Chemistry is a paperback series that offers selected thematic volumes from Springer chemistry series to graduate students and individual scientists in industry and academia at very affordable prices. Each volume presents an area of high current interest to a broad non-specialist audience, starting at the graduate student level.

Formerly published as hardcover edition in the review series
Structure and Bonding (Vol. 89) ISBN 3-540-62874-6

Cataloging-in-Publication Data applied for

ISBN 3-540-65553-0
Springer-Verlag Berlin Heidelberg New York

Cover: design & production, Heidelberg
Typesetting: Medio, V. Leins, Berlin
SPIN: 10711970   02/3020 - 5 4 3 2 1 0 - Printed on acid-free paper

Preface

This is the second of 3 special volumes of Structure and Bonding (88, 89 and 90) on recent advances in inorganic biochemistry. It is fitting that this volume begins by two chapters on zinc; one, by Kimura, Koike and Shinoya on the fashioning of zinc complexes as fitting models for the structure of relevant sites in proteins and enzymes; the other, by Auld, which describes, authoritatively, the multitude of tasks undertaken by enzymes that are dependent on zinc. This provides a wonderful example of the development of a subject, acknowledged to be important, but rather neglected by inorganic chemists until now. How that has changed! – this was due to the work of B. L. Vallee and his colleagues, J. F. Riordan and D. S. Auld, who, slowly but surely, showed that more and more enzymes require zinc either at the active site or in a structural capacity (or, of course, both). Coupled with the successful application of X-ray crystallography and the introduction of EXAFS, our understanding of the role of zinc has improved rapidly. However, the subject was given enormous stimulus by the discovery that zinc was absolutely essential in the reading of DNA. Zinc fingers "attracted" much attention: who knows, we may soon see a function for the protein that was first isolated in the fifties, metallothionein: its days as a Cinderella may soon be over!

Though it is hard to imagine vanadium playing such a crucial role, there has been much work over the past decade on its function, once doubted but now confirmed, in biology. Slebodnick, Hamstra and Pecoraro bring the whole panoply of spectroscopic techniques to bear on relevant vanadium complexes, and its biological role is elegantly addressed by Butler. The role of metal ions as Lewis acids and its implication in enzyme-catalysed phosphate monoester hydrolysis is authoritatively described by Gani and Wilkie. They touch on the "purple acid phosphatase": this enzyme is considered more extensively in the chapter by Klabunde and Krebs. One cannot fail to be impressed at the economy of Nature in the sense that, whilst retaining a dimeric metal ion core to the protein, the replacement of one of iron ions by zinc, leads to the phosphatase activity. (The properties of enzymes that contain di-iron centres are described by Sjöberg in Volume 88).

We are grateful to the authors for bringing us up-to-date on these important topics and hope that the articles will both inform and entertain the reader.

H. Allen O. Hill, Peter J. Sadler and Andrew J. Thomson

Contents

Advances in Zinc Enzyme Models by Small, Mononuclear Zinc(II) Complexes
E. Kimura, T. Koike, M. Shionoya   1

Zinc Catalysis in Metalloproteases
D. S. Auld 29

Modeling the Biological Chemistry of Vanadium: Structural and Reactivity Studies Elucidating Biological Function
C. Slebodnick, B. J. Hamstra, V. L. Pecoraro 51

Vanadium Bromoperoxidase and Functional Mimics
A. Butler, A. H. Baldwin 109

Metal Ions in the Mechanism of Enzyme-Catalysed Phosphate Monoester Hydrolyses
D. Gani, J. Wilkie 133

The Dimetal Center in Purple Acid Phosphatases
T. Klabunde, B. Krebs 177

Advances in Zinc Enzyme Models by Small, Mononuclear Zinc(II) Complexes

Eiichi Kimura,[1] Tohru Koike[1] and Mitsuhiko Shionoya[2]

[1] Department of Medicinal Chemistry, School of Medicine, Hiroshima University, Kasumi 1-2-3, Minami-ku, Hiroshima, 734, Japan. *E-mail: ekimura@ipc.hiroshima-u.ac.jp*
[2] Institute for Molecular Science, Okazaki National Institutes, Nishigonaka 38, Myodaiji, Okazaki, 444, Japan

Recent developments in zinc enzyme models, in particular for carbonic anhydrase (CA), alkaline phosphatase (AP), and alcohol dehydrogenase (ADH) are presented. Although these models are simple zinc(II) complexes, they have helped to disclose the hitherto unsettled intrinsic properties of zinc(II)-dependent enzyme functions. The discussion emphasizes how H_2O is activated by zinc(II) for the nucleophilic attack on electrophilic substrates (e.g., CO_2 in CA, phosphomonoesters in AP) and also how alcohols are activated by zinc(II) for hydride transfer in ADH or nucleophilic attack on phosphates in AP. Future modeling should take into consideration the results from recent developments in enzyme functions by protein engineering. For instance, design of secondary zinc(II) ligands for the fine-tuning of zinc(II) properties will be needed to explore and understand the reaction specificity of zinc enzymes.

Keywords: Zinc enzyme models, carbonic anhydrase, alkaline phosphatase, alcohol dehydrogenase, zinc(II) complexes, nucleophilic attack.

1	Introduction	2
2	Zinc Enzyme Models	2
2.1	Carbonic Anhydrase Model with Macrocyclic Triamine [12]aneN$_3$	2
2.2	Carbonic Anhydrase Models with Other Complexes	5
2.3	Model Study of Basicity of Zinc(II)-Bound OH$^-$ in Carbonic Anhydrase	7
2.4	Activation of Alcohols by Zinc(II) Ion in Alcohol Dehydrogenase	11
2.5	Activation of Proximate (Intramolecular) Alcohols by Zinc(II) Ion for Nucleophilic Reactions	15
2.6	Activation of Alcohols with Other Metal Ions	22
3	Recent Progress in Protein Engineering of Carbonic Anhydrase and Alkaline Phosphatase and Future Modeling	24
4	References	27

Structure and Bonding, Vol. 89
© Springer Verlag Berlin Heidelberg 1997

1
Introduction

Zinc(II) ion is a biologically essential element. Recognition of its importance is ever increasing, as more and more enzymes are discovered which contain zinc(II) in their active center. We have previously reviewed the fundamental properties of the zinc(II) ion including its Lewis acidity and the basicity of zinc(II)-bound OH^- using our zinc enzyme models (e.g., macrocyclic poly-amine zinc(II) complexes) [1–4]. More recent model studies by us and a large number of other groups have been establishing a deeper and wider scope of knowledge about the zinc(II) ion in relation to its biological functions. In this review, we wish to present the latest results obtained with recently designed zinc enzyme models.

2
Zinc Enzyme Models

2.1
Carbonic Anhydrase Model with Macrocyclic Triamine [12]aneN₃

Until now, several kinds of mononuclear zinc(II) complexes (e.g., 1 [1–10], 2 [1–4, 11–14], 3 [15–18], 4 [19–22], and 5 [23]) have been designed to mimic the zinc(II)-coordination structure or function of the zinc(II) ion at the active center of carbonic anhydrase (CA), the enzyme which catalyzes CO_2 hydration ($CO_2 + H_2O \rightarrow HCO_3^- + H^+$) and its reverse reaction HCO_3^- dehydration ($HCO_3^- \rightarrow CO_2 + OH^-$) [24].

However, the 12-membered macrocyclic triamine ([12]aneN₃) zinc(II) complex 1 has for the first time provided a convincing chemical mechanism which shows the role of zinc(II) in the reversible CO_2 hydration and HCO_3^- dehydration catalyzed by CA[7]. The fast kinetics of the CO_2 hydration catalyzed by 1 was followed by H^+ production ($CO_2 + H_2O \rightarrow HCO_3^- + H^+$) at 25 °C, which was detected by using a pH indicator in buffer solution (pH 6–10 and a similar pK_a as that of the pH indicator). The kinetics in fact demonstrated the catalytic nature of the zinc(II) complex 1 at various pHs. A plot of the initial rates against total zinc(II) complex concentrations (= $[1]_{total}$) indicated that the CO_2 hydration rate varied linearly with $[1]_{total}$ and $[CO_2]$ to give an observed second-order rate constant $(k^h_{cat})_{obs}$. The $(k^h_{cat})_{obs}$ data are plotted as a function of pH in Fig. 1a. The sigmoidal curve is characteristic of a kinetic process controlled by an acid-base equilibrium and exhibits an inflection point (pK_{kin}) at about pH 7.4, which is almost the same as the potentiometrically determined pK_a value of 7.3 for 1a ⇌ 1b + H^+ [5]. Thus, 1b must be the genuine active species in the catalytic hydration of CO_2. The sigmoidal pH dependence for the CO_2 hydration with CA (although its pK_a is 6.9) [24] can thus be accounted for by a hydroxo complex zinc(II)-OH^- at the active center of CA reacting with CO_2. The rate law for the suggested CO_2 hydration mechanism is more precisely given by $(k^h_{cat})_{obs} = k^h_{cat} \cdot K_a/([H^+] + K_a)$, and from this were calculated the kinetically obtained pK_a value of 7.4 and the k^h_{cat} value of $6 \times 10^2 \, M^{-1} s^{-1}$ at 25 °C. The CO_2 hydration me-

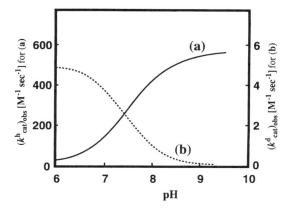

Fig. 1. The rate-pH profile for **a** CO_2 hydration and **b** HCO_3^- dehydration catalyzed by zinc(II)-[12]aneN$_3$ **1**

chanism for **1** is shown by reaction (1). The much faster reaction of CA (e.g., k_{cat} = ca. 10^8 M^{-1} s^{-1} for human CA II at 25 °C [25]) in comparison to **1** may be partially explained by an effective preassociation of CO_2 within the hydrophobic pocket in CA [26]. If this preassociation of CO_2 is characterized by a large binding constant of 10^5 M^{-1}, it can account for the large difference observed in the catalytic activity during the hydration of CO_2. In the real enzyme reaction, the rate-determining step is the proton transfer to solvent involving the nearby His(64) as a proton shuttle.

$$CO_2 + ZnL\text{-}OH^- \text{ (1b)} \xrightarrow[\text{slow}]{k^h_{cat}} [ZnL\text{-}^-OCO_2H] \xrightarrow{\text{fast}} HCO_3^- + ZnL\text{-}OH_2 \text{ (1a)} \quad (1)$$

The kinetic study of the reverse HCO_3^- dehydration catalyzed by a model complex was successfully conducted for the first time with the zinc(II) complex **1** at 25 °C [7]. The rate was followed by measuring the evolution of OH^- for the reaction ($HCO_3^- \rightarrow CO_2 + OH^-$) in a similar fashion to the CO_2 hydration, using a pH-indicator in buffer (pH 6–9) solution. The rates increased with lowering pH (see Fig. 1b). The kinetic data were fitted to the reaction (2), where the kinetically reactive species was **1a** and the second-order rate was followed with [**1a**] and [HCO_3^-], each first-order dependent. The dehydration constant k^d_{cat} with **1a** was found to be 5 M^{-1} s^{-1} and the kinetically obtained pK_a value for **1a** $\Leftrightarrow$ **1b** + H^+ was 7.3 at 25 °C. Important conclusions for HCO_3^- dehydration with the catalyst **1** are (i) the reactive species is the zinc(II)-OH_2 form, (ii) substitution of the zinc(II)-bound H_2O with HCO_3^- is rate determining, and (iii) decarboxylation of the zinc(II)-bound HCO_3^- is much faster.

$$HCO_3^- + ZnL\text{-}OH_2 \text{ (1a)} \xrightarrow[\text{slow}]{k^d_{cat}} [ZnL\text{-}^-OCO_2H] \xrightarrow{\text{fast}} CO_2 + ZnL\text{-}OH^- \text{ (1b)} \quad (2)$$

Although the magnitudes of k^h_{cat} and k^d_{cat} differ significantly, the two curves (Fig. 1a and b) have an inflection point at the same pH (ca. 7.4) and are sym-

1a: X= H₂O   2a: X= H₂O
b: X= OH⁻   b: X= OH⁻   3

4   5

metrical (with scales adjusted). Thus, the model complex 1 is the first example to mimic the pH-dependent behavior of reversible CO_2 hydration catalyzed by CA. This fact implies that in CA, too, the CO_2 hydration/HCO_3^- dehydration is determined by the zinc(II)-OH⁻/zinc(II)-OH_2 equilibrium at the active center. A similar CA model study with 2 recently gave an analogous reaction mechanism [14].

In all the other model complexes [27, 28] either only CO_2 uptake or decarboxylation (with Co^{3+}, Rh^{3+}, Ir^{3+}, Cr^{3+}), or only HCO_3^- substitution (with Pd^{2+}) was observed. The existence of pH-control and reversibility for the latter complexes was unknown. The zinc(II) complex 1b is significantly more reactive ($k_{cat}^h = 6 \times 10^2$ M⁻¹s⁻¹) than Woolley's model complex 5b ($k_{cat}^h = 2 \times 10^2$ M⁻¹s⁻¹) [23]. In another model, the tris(imidazole) complexes 3, higher rate constants of HCO_3^- dehydration ($k_{cat}^d = 900 - 2800$ M⁻¹ s⁻¹) were reported in 80% EtOH/H_2O at pH ca. 6.5 [18]. However, it failed to mimic the characteristic pH profile for both the hydration and dehydration reaction of CA. The model complex 1 demonstrated the unique properties of a labile water molecule that is susceptible to substitution reactions with the substrate HCO_3^-, anion inhibitors, and a non-labile OH⁻ ligand that acts as a nucleophile and attacks CO_2. By changing the pH (around a physiological pH of ca. 7) zinc(II) can choose either reactant. The zinc(II)-bound HCO_3^- has the choice either to be substituted by OH⁻ for the reverse aquation reaction at alkaline pH, or to lose CO_2 for the decarboxylation at acidic pH. In this context it is of great significance that the bicarbonate anion (HCO_3^-) has the highest 1:1 affinity constant ($K =$ [ZnL-⁻OCO₂H]/[ZnL][HCO_3^-] = $10^{4.0}$ at pH 8.4) for 1a, second only to hydroxide anion ($K = 10^{6.4}$) (see Table 1) [9]. Taken together, it can be concluded that zinc(II) with a pK_a value of ca. 7 for the zinc(II)-bound H_2O is probably the most appropriate metal ion to perform the functions of CA at a physiological pH of ca. 7. Also the very high affinity of zinc(II) for HCO_3^- that binds as a

Table 1. A comparison of anion affinity constants (log K)[a] for **1a** at 25 °C

anion	log K	
OH⁻	6.4[b]	
HCO₃⁻		4.0[c]
CH₃COO⁻	2.6,[b]	2.5[c]
SCN⁻	2.4,[b]	2.0[c]
Br⁻	1.5,[b]	1.5[c]
Cl⁻	1.3,[b]	1.5[c]

[a] $K = [\text{ZnL-anion}]/[\text{ZnL}][\text{anion}]$ (M⁻¹).
[b] Determined by potentiometric pH titration.
[c] Determined by 4-nitrophenyl acetate hydrolysis kinetics.

monodentate ligand is a critical property of zinc(II). If the metal were more acidic and/or had higher coordination numbers (e.g., cobalt(III)), HCO₃⁻ may be deprotonated to CO_3^{2-} to act as a bidentate ligand, and would no longer be susceptible to decarboxylation.

Although **1** offers an excellent CA model, its catalytic activity is very moderate in comparison to CA. Model **1** simply illustrates the essence of the intrinsic properties of zinc(II) at the active center of CA. Other important features, such as the hydrophobic pocket (for CO_2 entry), proton relay (network), and other structures in CA [26], are missing (see Sect. 3). More sophisticated next generation models should be equipped with these functions.

2.2
Other Carbonic Anhydrase Models

A novel tris(pyrazolyl)borate zinc(II) complex **4b** (R = p-isopropylphenyl, R' = Me, X = OH⁻) was prepared by mixing zinc(II), ligand and KOH in MeOH/CH₂Cl₂ [29]. Due to insolubility or instability of the zinc(II) complex in aqueous solution, the pK_a value for the zinc(II)-bound H₂O in aqueous solution was not reported. The zinc(II)-OH⁻ in **4b** showed sufficient nucleophilicity towards hydrolysable substrates, carboxylesters, activated amides (β-lactam or CF₃CONH₂), and phosphonates in CH₂Cl₂ or benzene solution (see reactions 3–5), where the reactions are not catalytic, but stoichiometric. The seemingly high nucleophilicity of the tris(pyrazolyl)borate zinc(II) complex was attributed to steric hindrance and the hydrophobic environment around the zinc(II)-bound OH⁻. It is of interest to see whether **4b** acts as a catalyst in H₂O-containing solvents. The zinc(II)-OH⁻ species **1b** and **2b** catalytically hydrolyze esters [5], β-lactams [11], and bis(4-nitrophenyl) phosphodiester in aqueous solution [6], although they were not tested with phosphonates.

$$\text{RCOOR}' + \text{ZnL-OH}^- \rightarrow \text{ZnL-}^-\text{OCOR} + \text{R}'\text{OH} \tag{3}$$
$$\text{RCONHR}' + \text{ZnL-OH}^- \rightarrow \text{ZnL-}^-\text{OCOR} + \text{R}'\text{NH}_2 \tag{4}$$
$$\text{HPO(OMe)}_2 + \text{ZnL-OH}^- \rightarrow \text{ZnL-}^-\text{OPHOMe} + \text{MeOH} \tag{5}$$

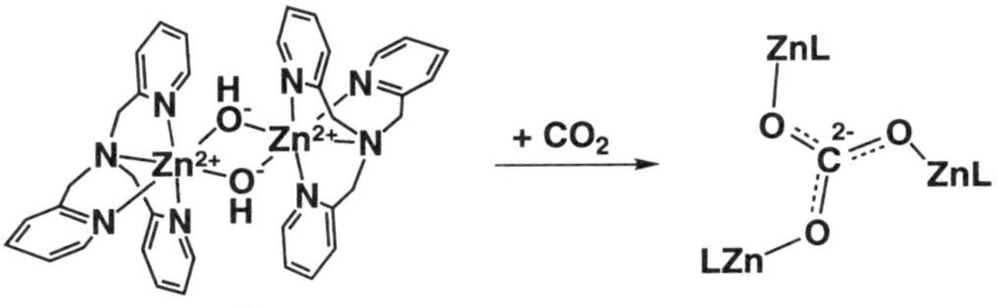

6

LZn(μ-OH)$_2$ZnL

7

A new di-μ-hydroxo di-zinc(II) complex **6** was prepared by treatment of tris(2-pyridylmethyl)amine and zinc(II) perchlorate in MeOH with 1 equiv KOH [30]. When a solution (in MeOH or H$_2$O) of **6** is exposed to air, CO$_2$ is readily attacked by the zinc(II)-OH$^-$ species to form a stable carbonato complex **7**. The structure has been elucidated by X-ray crystal analysis.

As a structural model of the zinc(II)-OH$^-$ species of carbonic anhydrase (CA), the first mononuclear tetrahedral (HO$^-$-bound) zinc(II) complex **4b** (R = t-Bu, R′ = Me) was prepared by mixing zinc(II), the ligand and KOH, and characterized by X-ray crystal analysis [19]. The zinc(II)-OH$^-$ complex **1b** isolated earlier was a cyclic trimer linked by three hydrogen bonds between each zinc(II)-bound hydroxide group, as shown by an X-ray crystal study [5]. However, in aqueous solution, this dissociates into monomeric zinc(II)-OH$^-$ species. The zinc(II) complex **4b** in CHCl$_3$ solution reacted immediately with CO$_2$, possibly to form a HCO$_3^-$ complex **8** that ultimately and irreversibly gave a bridging carbonato complex **9**. However, quantitative determination of the pK_a value for **4a** $\rightleftarrows$ **4b** + H$^+$ or of the nucleophilicity of **4b** was not reported. The HCO$_3^-$ complex **8** was characterized by IR spectroscopy (1675 and 1302 cm^{-1} for zinc(II)-bound bicarbonate) [20]. The formation of the bicarbonate complex **8** is reversible, and removal of the CO$_2$ atmosphere results in regeneration of **4b**,

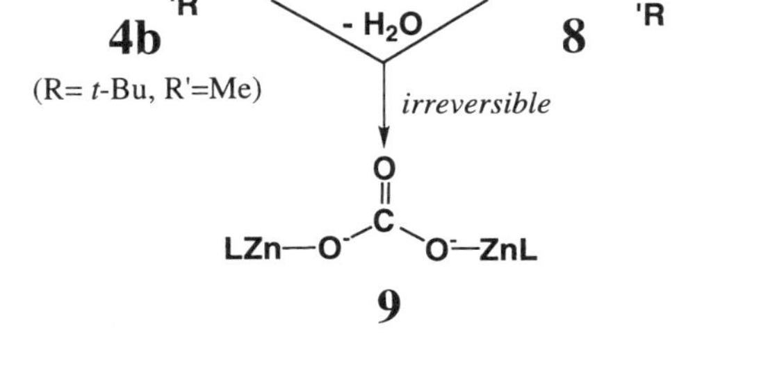

4b

(R= t-Bu, R′=Me)

8

irreversible

9

as detected by ^{1}H NMR spectroscopy. In view of the labile equilibrium between **4b** and **8**, attempts to crystallize **8** were made under CO_2. However, a symmetrically bridging carbonato complex **9** was isolated instead. The hydroxide complex **4b** effectively promoted the exchange of oxygen atoms between $CO_2 + H_2^{17}O$, which served as a good functional model for CA. The authors suggested that the facile equilibrium between zinc(II)-OH$^-$ and zinc(II)$^-$OCO$_2$H may be a critical function for CA activity. The increased tendency towards bidentate coordination (due to stronger acidity) across the metal series ($Zn^{2+} < Co^{2+} < Ni^{2+}$ and Cu^{2+}) was observed, which resulted from stronger binding of the bicarbonate ligand. The authors concluded that in this context, too, zinc(II) is probably the best suited metal for CA.

Recently, a new tris(imidazolyl)phosphine zinc(II) complex **10**, where X is a monodentate ligand (= OH$^-$, I$^-$, and NO$_3^-$), was synthesized and examined by X-ray crystal analysis [31]. The zinc(II)-bound OH$^-$ complex is claimed to be the first structurally characterized monomeric zinc(II) hydroxide complex sequestered by three imidazole groups and may therefore become an excellent structural model for the active site of carbonic anhydrase.

All of these recent results qualitatively or quantitatively illustrate the importance of nucleophilicity of zinc(II)-bound OH$^-$ species in aqueous or non-aqueous solutions. This aspect is extremely important in almost all of the zinc enzymes and will be presented again for models of other hydrolytic enzymes. However, in CA and many other zinc enzymes, the basicity of the zinc(II)-bound hydroxide is also extremely important.

2.3
Model Study of Basicity of Zinc(II)-Bound OH$^-$ in Carbonic Anhydrase

Aromatic sulfonamides are strong inhibitors of carbonic anhydrase (CA), among which acetazolamide ($K_D = 6.0$ nM for human carbonic anhydrase II (HCA II) at pH 7.4 [24]) is therapeutically prescribed as a diuretic drug. The question as to how acetazolamide is bound to the active center appears to have been settled by X-ray crystal studies (see Fig. 2), although the precise inner-sphere binding mode of the sulfonamide to zinc(II) is not completely solved. The focal point has been whether or not the sulfonamides are deprotonated. The crystal X-ray resolutions failed to afford a clear-cut answer [32]. The alcoholic O atom of Thr(199) is linked by a hydrogen bond with the sulfonamide proton, possibly assisting the easy deprotonation. Chemically, it is puzzling that

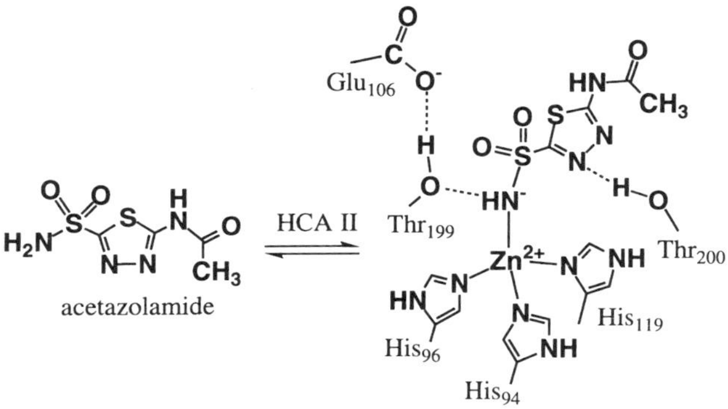

Fig. 2. A proposed acetazolamide binding mode at the carbonic anhydrase active center

undeprotonated sulfonamides ($pK_a = 7-10$) can be good donors, forming strong coordination bonds with the zinc(II) ion. One of the pieces of indirect evidence for the deprotonation is that the lower the pK_a values of the sulfonamides are (e.g., pK_a of 7.4 for acetazolamide), the higher the 1:1 affinity constants. On the basis of formation of a strong fluorescent complex of CA with dansylamide ($pK_a = 9.9$ for the sulfonamide, $K_D = 1.1$ μM with HCA II at pH 7.4 [33]), it was concluded that the deprotonated form of dansylamide binds to zinc(II) at the CA active center. In any case, an interaction of the aromatic group of dansylamide with the protein must make a large contribution to the observed stability of the complex. A question concerning the intrinsic zinc(II) properties was whether it is acidic enough to displace the proton of sulfonamides, or whether the zinc(II)-bound OH^- (generated at physiological pH) is basic enough to neutralize the weak acid sulfonamide group? However, these fundamental questions remain unanswered. Only model studies would give an insight into the interaction of aromatic sulfonamides with zinc(II). The zinc(II) complexes with [12]aneN$_3$ (1) and [12]aneN$_4$ (2) were to our knowledge the first models that were undertaken for that purpose.

Treatment of a typical sulfonamide CA inhibitor, acetazolamide, with **1a** immediately yields the product as a precipitate, which was identified as a 1:1 complex **11**, wherein the sulfonamide is deprotonated and coordinated to the zinc(II) ions [5]. It is of interest to point out that with other transition metal ions (e.g., Ni^{2+}) acetazolamide binds to the thiazole N, but not to the deprotonated sulfonamide nitrogen [34]. This different behavior of zinc(II) best illustrates its outstanding (hard) acid nature that favors the anionic N^- donor over the neutral (soft) N donor. The pK_a value of 7.4 for deprotonation of acetazolamide is close to that for the zinc(II)-bound H_2O in 1, which would be extremely favorable for simultaneous coordination and deprotonation to yield the 1:1 complex **11**. Aromatic sulfonamides with higher pK_a values are more difficult to deprotonate at neutral pH and hence the apparent 1:1 affinity with zinc(II) is weaker. More convincing evidence of the basicity of [12]aneN$_3$-zinc(II)-OH$^-$ was obtained from the interaction of the tosylamidopropyl-pendant [12]aneN$_3$ 12 (HL) with the zinc(II) ion [9]. A stable amide-deprotonated complex **13**

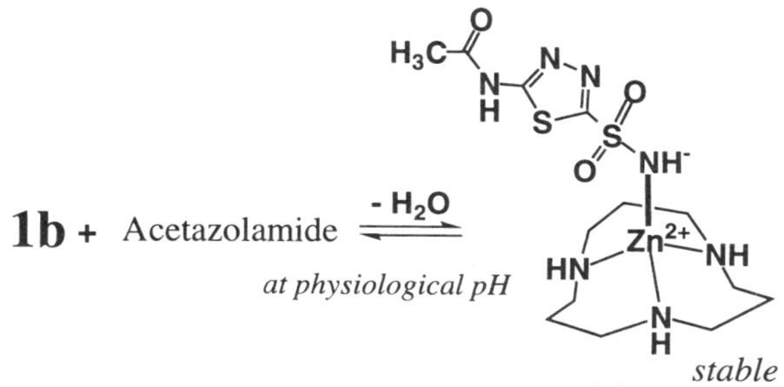

11

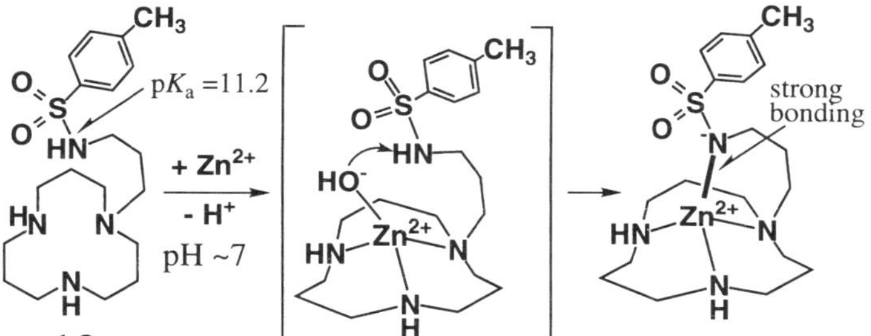

12
(HL)

13
(ZnL)

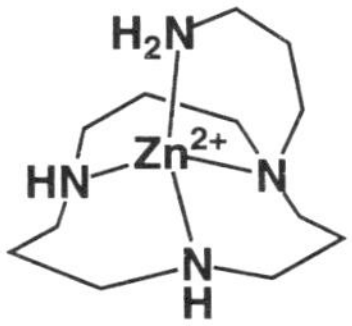

14

(ZnL) was formed at physiological pH, despite the pK_a of the tosylamide being 11.2. The structure of **13** was determined by an X-ray crystal analysis. The comparison of the complex stability for **13** (log K = [ZnL]/[Zn][L] = 14.7) with those of **1a** (log K = 8.4) and **14** (log K = 11.7) indicates that the zinc(II) ion in the triamine complex prefers the fourth ligand in the order of: anionic nitrogen > neutral nitrogen > water. The complex **11** has lost the ability to catalyze ester hydrolysis. Thus, the novel zinc(II) complex **11** may chemically represent the sulfonamide inhibition of CA. On the basis of this model study, we consider that the zinc(II) coordinated by three histidyl imidazoles at the active center of CA is sufficiently acidic to deprotonate the aromatic sulfonamides, especially acetazolamide. Other neutral organic compounds, such as phenols and carboxylic acids, are also inhibitors of CA. Similar acid-base properties should operate in their interactions.

Very recently, dansylamidoethyl-pendant cyclen 15 (HL) has been synthesized as a model for the CA-dansylamide complex [35]. The 1:1 zinc(II) complex 16 (ZnL) was shown again to contain the zinc(II)-bound sulfonamide N^- anion by X-ray crystal analysis. The comparison of 1:1 complexation constants for 2a (log K = 15.3) and 16 (log K = 20.8) indicates that the extra $10^{5.5}$ fold stability comes from the sulfonamide anion coordination. The dansylamidoethyl-pendant cyclen 15 is a novel type of zinc(II)-fluorophore [35]. 15 forms very stable complexes with zinc(II), cadmium(II) and copper(II) under physiological conditions. The 1:1 zinc(II) complex 16 shows a fluorescent maximum at 528 nm (quantum yield = 0.11) in aqueous solution, while the free ligand shows a weaker fluorescence at 555 nm (quantum yield = 0.03). The copper(II) complex, on the contrary, completely quenches the fluorescence. The zinc(II)-dependent fluorescence is quantitatively responsive over a 0.1–5 µM concentration range with 5 µM 15, and is not affected by the presence of millimolar concentrations of biologically important metal ions such as Na^+, K^+, Mg^{2+}, and Ca^{2+}. This new ligand 15, that forms a far more stable 1:1 zinc(II) complex than any previous zinc(II)-fluorophore, may be promising as a new zinc(II)-fluorophore.

15
(HL)

16
(ZnL)

From the kinetic study of 1b-catalyzed hydrolysis of 4-nitrophenyl acetate inhibited by various aromatic sulfonamides, the apparent 1:1 affinity constants were determined at pH 8.4 [9] (see Table 2). A comparison of intramolecular (K_{intra} = $10^{6.3}$ = K(for 13)/K(for 1a)) and intermolecular (K = $10^{2.4}$ M^{-1} for 1a with p-toluenesulfonamide) contributions of p-toluenesulfonamide anion coordination to the zinc(II)-[12]aneN$_3$ complex gives an effective molarity of K_{intra}/K = $10^{3.9}$ M for the intramolecular location. These 1:1 anion complexation constants for the zinc(II) macrocyclic triamine complexes with various aromatic sulfonamides exhibit a trend paralleling that reported for CA [36], suggesting a similar inhibition mechanism (i.e., the deprotonated sulfonamide anions can coordinate to the zinc(II) in the CA active center). The high effective molarity of the pendant sulfonamide in the model complex may be viewed as the non-coordinating contribution which brings the inhibitors to the zinc(II) center of CA.

To return to the question of the basicity of zinc(II)-bound OH^-, we can conclude that despite 1b (pK_a = 7.3 for its conjugate acid 1a) being a weaker base

Table 2. A comparison of sulfonamide anion affinity constants (log K)[a] determined from 4-nitrophenyl acetate hydrolysis kinetics at 25 °C

	log K	
	1a	carbonic anhydrase[b]
acetazolamide	3.6	7.9
4-nitrobenzenesulfonamide	2.6	7.2
p-toluenesulfonamide	2.4	6.3

[a] $K = [\text{ZnL-sulfonamide}^-]/[\text{ZnL}][\text{unbound sulfonamide}]$ (M^{-1}).
[b] From ref. 36.

than aromatic sulfonamides with a higher pK_a (e.g., $pK_a = 10.5$ for p-toluene-sulfonamide), zinc(II)-OH$^-$ can still deprotonate the aromatic sulfonamides, because the resulting sulfonamide anions ($ArSO_2NH^-$) have extremely strong interactions with zinc(II), compensating for the unfavorable interaction between the weak base and the weak acid. It may be inferred that while NaOH releases OH$^-$ anion without any favorable contribution from Na$^+$, ZnL-OH$^-$ releases OH$^-$ with ZnL^{2+} waiting to accept the conjugate base.

2.4
Activation of Alcohols by Zinc(II) Ion in Alcohol Dehydrogenase

Zinc(II)-containing alcohol dehydrogenases (ADH) catalyze hydride transfer from alcohols to NAD$^+$. The X-ray crystal structure of horse liver ADH reveals that the active site of the enzyme contains tetrahedral zinc(II) coordinated with two thiolates from Cys(46) and Cys(174), one imidazole of His(67) and a water molecule [37]. Studies performed with the enzyme led to the hypothesis that, in the course of the proton transfer starting from His(61), the acidic zinc(II) simultaneously (directly or indirectly) assists generation of an alkoxide ion from the substrate alcohol at physiological pH (see Fig. 3). However, the question may be raised as to whether the high pK_a values of alcohols (normally ca. 16) can be lowered sufficiently to yield an alkoxide anion in the vicinity of zinc(II). Alcohols in the inner coordination sphere of zinc(II) can be protonated or deprotonated (alkoxide), as shown by the model complexes, see below. In Sect. 2.1, we showed in model studies that zinc(II)-bound H_2O can have decreased pK_a values of ca. 7. Meanwhile, in ADH the pK_a of the H_2O bound to zinc(II) varies dramatically with the substrate-bound ternary complexes from 9.2 in the free form, 11.2 in NADH-bound form and 7.6 in NAD$^+$-bound form [38]. Moreover, when the zinc(II)-OH$_2$ in the ternary complex with NAD$^+$ is replaced by zinc(II)-alcohol, the pK_a is lowered to 6.4. It would be of interest to test whether [12]aneN$_3$-zinc(II)-OH$_2$ **1a** with a pK_a value of 7.3 (similar to 7.6 in the NAD$^+$-bound ADH) would catalyze hydride transfer from an alcohol to an aldehyde or an NAD$^+$ model [39]. In this model, the electrostatic effect of NAD$^+$ would hardly operate due to intermolecular interaction with the catalytic center.

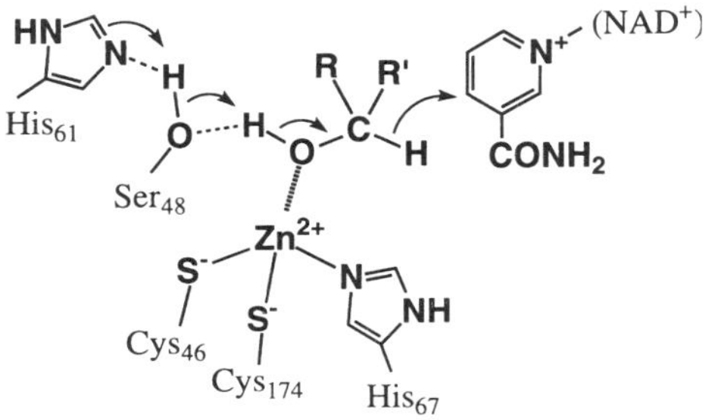

Fig. 3. A proposed "hydride-transfer" mechanism at the active center of alcohol dehydrogenase (ADH)

When 0.125 mmol 4-nitrobenzaldehyde (**17**) was heated at reflux for 24 h under an argon atmosphere in 1.0 mL of an alcohol (2-propanol, 2-propanol-d_8, EtOH or MeOH) containing 0.8 mol% of a zinc(II) complex (i.e., 1 mM of $(1b)_3(CF_3SO_3)_3 \cdot CF_3SO_3H$, $2a \cdot 2ClO_4^-$ or other zinc(II) salts) [39], the reaction gave only two products, 4-nitrobenzyl alcohol (**18**) and 4-nitrobenzaldehyde dialkyl acetals (**19**). Comparison of the product distribution and yields with various zinc(II) species is highly instructive.

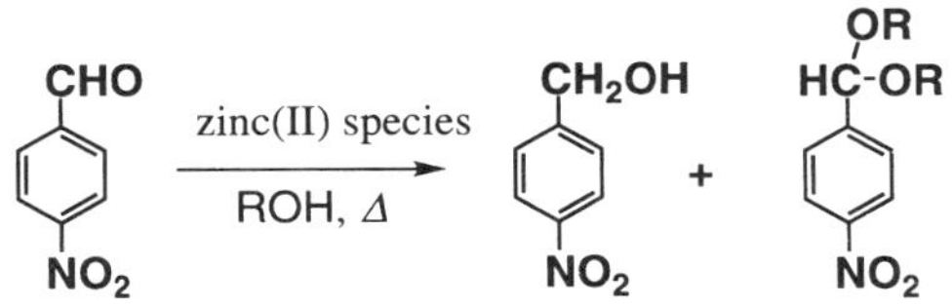

Most significantly, the reaction in 2-propanol with $(1b)_3(CF_3SO_3)_3 \cdot CF_3SO_3H$ after 24 h gave the product in 7820% catalytic yield (corresponding to 62.4% consumption of the starting aldehyde). Other zinc(II) species were virtually non-catalytic in this ADH-mimicking reaction. The reaction in 2-propanol-d_8 catalyzed by $(1b)_3(CF_3SO_3)_3 \cdot CF_3SO_3H$ (see Fig. 4) was monitored by ^{1}H NMR for 40 h, which unequivocally proved the quantitative "D$^-$" transfer to yield a monodeuterated 4-nitrobenzyl alcohol **20** at the benzylic position with a small amount of the corresponding acetal **21** (370% catalytic yield). The peak integration ratio of the benzylic H proton over the four aromatic protons was constant (= 1:4) during this reaction, indicating no reverse hydride transfer form of the product.

The outstanding yield of hydride-transferred product **20** by $(1b)_3(CF_3SO_3)_3 \cdot CF_3SO_3H$ over other zinc(II) species strongly suggests that the most acidic zinc(II) in the zinc(II)-[12]aneN$_3$ complex transiently generates zinc(II)-alkoxide (by attacking the zinc(II)-OH$^-$) and the α C-H bond is then labilised. In view of the low barrier for tetra- to pentacoordinate configurational intercon-

$$17 \quad \underset{\underset{\Delta}{2\text{-PrOH-}d_8}}{\overset{1b}{\underset{\times}{\rightleftarrows}}} \quad \overset{\overset{D}{\underset{C\text{-OH}}{|}}}{\underset{NO_2}{\bigcirc}} \quad + \quad \overset{\overset{OPr}{\underset{HC\text{-OPr}}{|}}}{\underset{NO_2}{\bigcirc}}$$

20   **21**

version with zinc(II)-[12]aneN₃, a mechanism comprising both tetra- and pentacoordination is proposed, as shown in Fig. 4. The hydroxide ion in **1b** may act as a base to generate a transient alkoxide complex. The zinc(II)-[12]aneN₃ complex, which has the largest open space available for the reaction, facilitates smooth hydride transfer. The mechanism above is nearly the same as the hydride transfer postulated for ADH in Fig. 3 [37], where alkoxide formation on zinc(II) was not necessarily invoked. As shown later, zinc(II)-bound alcohols show pK_a values of ca. 7, almost the same as those of zinc(II)-OH₂. Thus, the postulated low pK_a of 6.4 assigned to ADH-alcohol vs the pK_a of 7.6 for ADH-OH₂ [38] may not apply. The "Meerweine-Pondorf-Varley" reaction using aluminum(III) triisopropoxide, Al(Me₂CHO)₃, involves a similar hydride transfer from isopropoxide to the carbonyl group on aluminum(III) (see Fig. 5). Here, the product RR'CHO⁻ is thought to be strongly bound to aluminum(III), since the latter is an extremely strong acid. Accordingly, the reverse hydride

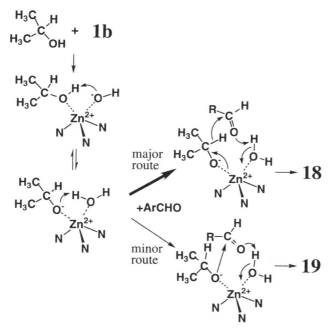

Fig. 4. The zinc(II)-[12]aneN₃-catalyzed hydride-transfer reaction

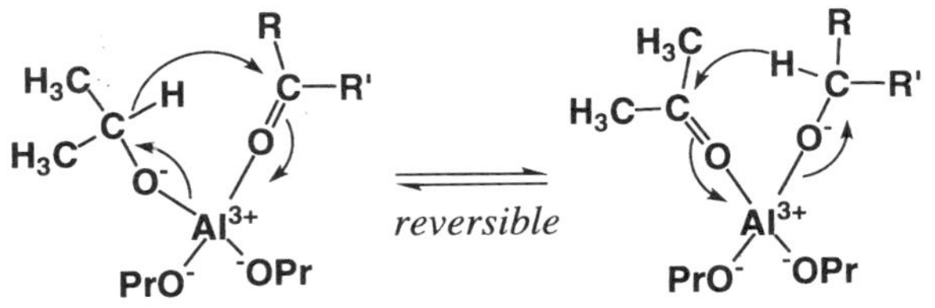

Fig. 5. An aluminum(III)-catalyzed hydride transfer-reaction

transfer from the product alcohol to acetone can also occur. In contrast, the zinc(II) species is a milder acid and the interaction between it and the alkoxide is weaker, so that smooth ligand exchange can occur on the zinc(II) ion. Consequently, the catalytic turnover of the hydride transfer is much greater.

A new model for the interaction of alcohol with the zinc(II)-bound thiolate site of alcohol dehydrogenase was recently reported by Kovacs et al. [40]. An X-ray crystal study of (1,12-dimethyl-3,7,11-triazatrideca-2,11-diene-1, 13-ditholato)zinc(II)-MeOH showed that the zinc(II) complex is pentacoordinate and trigonal bipyramidal. On the basis of this X-ray crystal structure, the authors speculated that the zinc(II)-bound thiolate anion in ADH may participate in the activation of substrate alcohols by forming a hydrogen bond between the S$^-$ and the alcoholic hydroxyl group to assist the alkoxide formation. Whether this is indeed the case should be tested by seeing whether this model complex is able to convert alcohols into aldehydes in the presence of the hydride acceptor molecules.

Interestingly, the zinc(II)-bound alkoxide derived from zinc(II)-[12]aneN$_3$ simultaneously catalyzed a nucleophilic reaction at the aldehyde, which corresponds to the hydration of RR'CO (or CO$_2$) in aqueous solution [39]. A similar type of reaction will be discussed in the next section. When the hydride acceptor is changed to the NAD$^+$ model **22** under the same conditions, almost exclusive formation of the 1,4-dihydro adduct **23** was observed, as is the case in the ADH reaction (see Fig. 6). No other product other than a minor 1,6-adduct **24** was detected. Interestingly, a reverse reaction (hydride transfer in the NADH model **23** to **22**) was not catalyzed by (**1b**)$_3$(CF$_3$SO$_3$)$_3$ · CF$_3$SO$_3$H. However, the

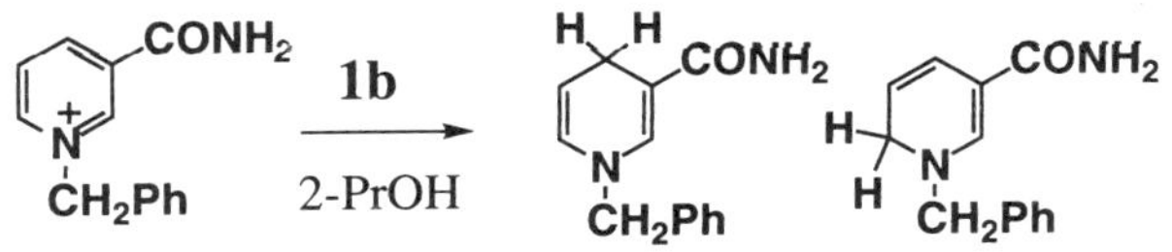

Fig. 6. The reduction of an NAD$^+$ model with **1b**

reverse reaction occurred with weaker acids such as $Mg(ClO_4)_2$ and $Zn(CF_3SO_3)_2$, which reflects the weaker acidic properties of zinc(II) ($pK_a = 11.2$) in the ternary complex of ADH with NADH. The model reaction of zinc(II)-[12]aneN$_3$ indicated that the substrate alcohols are certainly activated on acidic zinc(II). As to the mode of activation of substrate alcohols by zinc(II) in ADH, it was still not clear, for instance, whether ROH is bound to zinc(II) as RO$^-$ or ROH, or whether, in the presence of inert zinc(II)-OH$^-$ (with a pK_a of 7.6) alcohols may interact directly with zinc(II). Our model study seems to offer an answer to questions about the enzyme. The zinc(II)-OH$^-$ (with a pK_a of 7.3) can act as a base to attack incoming alcohols, yielding a transient alkoxide (partially stabilized by zinc(II)), which in turn gives the hydride transfer product (see Fig. 4).

2.5
Activation of Proximal (Intramolecular) Alcohols by Zinc(II) Ion for Nucleophilic Reactions

Activation of proximal (intramolecular-type) alcohols by metal coordination is illustrated by serine(102) under the influence of zinc(II) at the alkaline phosphatase active center, which is involved in phosphate hydrolysis [41]. On the basis of X-ray crystal and NMR analyses, it is now considered that the

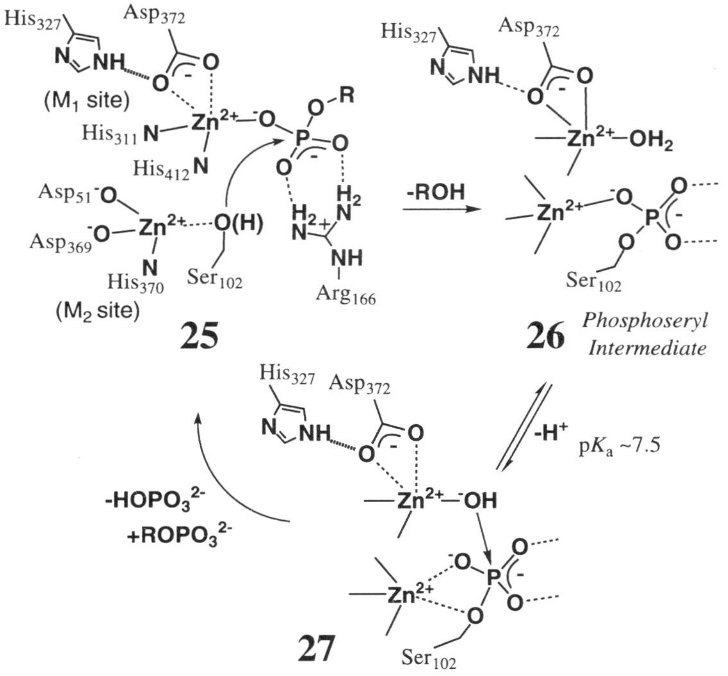

Fig. 7. Phosphomonoester hydrolysis at the active center of alkaline phosphatase.

phosphate substrate (bound to zinc(II) at the M1 site) is initially attacked by the zinc(II)-activated serine(102), alcohol or alkoxide (see the M2 site of **25** in Fig. 7) to yield a transient phosphoseryl intermediate **26**, which is then attacked intramolecularly by the adjacent zinc(II)-bound hydroxide (at the M1 site of **27**) to complete the hydrolysis and yield the free form of serine (102), thus reactivating the catalytic cycle. There are some interesting chemical questions which arise concerning the mechanism: (i) How does the serine hydroxyl group become a strong nucleophile upon association with the zinc(II) ion? and (ii) what is the special chemical advantage of forming the phosphoseryl inter-mediate **26** by such an indirect hydrolysis? The first question is relevant for the activation of substrate alcohols for the subsequent hydride transfer in the critical step of alcohol dehydrogenase (see Sect. 2.4).

In 1972, Sigman and Jorgensen studied a ternary complex of zinc(II)-N-(2-hydroxyethyl)ethylenediamine **28** and 4-nitrophenyl picolinate **29** as a catalytic model for zinc(II)-alkoxide-promoted transesterification to produce an ester intermediate **30** [42]. Indeed, they observed formation of the ester **30**. From the kinetic study it was concluded that the alcohol donor deprotonates to the alk-oxide with a pK_a of 8.4, which was an active nucleophile for ester bond cleavage of the substrate **29**. However, **30** was too stable and the subsequent hydrolysis, which is the last essential step to complete the catalytic cycle, was not observed. It is possible that the zinc(II) in **30** could not produce an active nucleophile, such as zinc(II)-OH$^-$, at experimental pH for the final hydrolysis. Despite this shortcoming, this was probably the first model that showed alcohols can be activated in the same way as H_2O on zinc(II), and that zinc(II)-bound alkoxide may be a stronger nucleophile than a zinc(II)-bound OH$^-$ ion.

The alkaline phosphatase reaction shares some common features with the well-known mechanism of serine enzymes such as chymotrypsin (see Fig. 8) [43]. However, a major difference between alkaline phosphatase and serine proteases is that the serine OH group is activated by zinc(II) in the former en-zyme, while in the latter it is activated by an adjacent histidyl imidazole, which in turn, is linked to a carboxylate anion. This suggests the question, "Why does nature use the zinc(II) ion in some cases and imidazole in others to activate the serine OH, generating the alcoholic nucleophile?".

As a continuation of zinc(II)-[12]aneN$_3$-catalyzed hydrolysis studies, an alcoholic pendant [12]aneN$_3$ **31** was synthesized to investigate whether the pendant alcohol, under the strong influence of the zinc(II) trapped in the macrocyclic ring, could be activated to become a strong nucleophile in a simi-

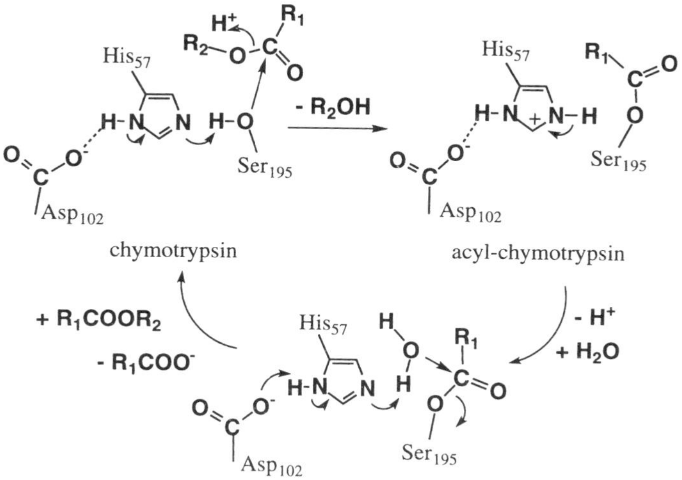

Fig. 8. The reaction mechanism for carboxyl ester hydrolysis by chymotrypsin.

lar fashion to the zinc(II)-activated serine OH in alkaline phosphatase [10]. Recently, the chymotrypsin model presented earlier [44] was investigated again. It was shown that this model does not work [45]. The potentiometric titration of **31** in the presence of zinc(II) showed that **31** yields a 1:1 zinc(II) complex **32a**, where the alcoholic OH deprotonates at physiological pH to form **32b**, which was isolated as a dimer and characterized by X-ray crystal analysis. In aqueous solution the zinc(II)-bound alkoxide in monomeric **32b** was found to be *a stronger nucleophile* (towards 4-nitrophenyl acetate) than the zinc(II)-bound OH$^-$ anion with the [12]aneN$_3$ complex **1b**. It is of interest that the water and the pendant alcohol at the fourth coordination site of zinc(II)-[12]aneN$_3$ deprotonates with almost the same pK_a values of 7.3 and 7.4, respectively, at 25 °C.

Although **32b** was initially designed for an alkaline phosphatase (AP) model, it reacted very slowly with phosphomonoester substrates of AP. Accordingly, the 4-nitrophenyl acetate hydrolysis catalyzed by **32b** was studied in an aqueous solution of 10 %(v/v) CH$_3$CN (see Fig. 9). The initial hydrolysis rate was followed by the appearance of a 4-nitrophenolate anion. The second-order rate constant k_{NA} obtained at 25 °C was 1.4×10^{-1} M^{-1} s^{-1}. This value did not change

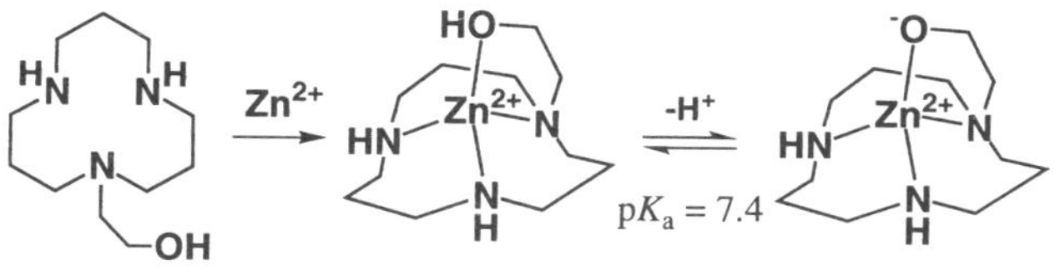

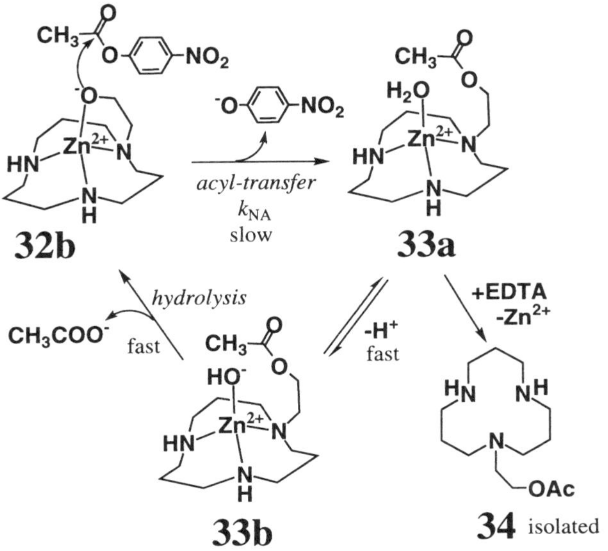

Fig. 9. An overall reaction mechanism for 4-nitrophenyl acetate hydrolysis catalyzed by an alcohol-pendant [12]aneN$_3$ zinc(II) complex

when D$_2$O was used. This negligible isotope effect suggests that the zinc(II)-bound alkoxide acts directly as a nucleophile, but not as a general base, producing a nucleophilic HO$^-$ anion (from the water molecule). Since the initial k_{NA} value held after more than one catalytic cycle, the hydrolysis by **32b** was concluded to be catalytic. The initial product was confirmed as a transient "acetyl intermediate" **33a** by isolation of the zinc(II)-free (by treatment of excess EDTA) ligand **34**. The zinc(II)-coordinating acetyl-intermediate **33a** was too reactive to be isolated. It rapidly underwent hydrolysis by the zinc(II)-bound OH$^-$ immediately generated(at the alkoxide-vacated site) in **33b**. Hence the rate-determining step in the overall reaction was the initial "acetyl transfer" process.

The second-order rate constant k_{NA} for 4-nitrophenyl acetate hydrolysis by **1b** under the same conditions was 3.6×10^{-2} M^{-1} s^{-1} at 25 °C [10]. Having both the kinetic data for the acetate hydrolysis catalyzed by the zinc(II)-bound alkoxide anion of **32b** and the zinc(II)-bound OH$^-$ of **1b**, we concluded that the zinc(II)-bound alkoxide is about 4 times as strong a nucleophile as the zinc(II)-bound OH$^-$ of **1b**, despite almost the same basicity for both anions. To our knowledge, this was the first demonstration that a zinc(II)-bound alkoxide can be a better nucleophile than a zinc(II)-bound OH$^-$, although the difference may not seem so remarkable. In the majority of the hydrolytic zinc(II) enzymes, a serine or threonine residue acts as a nucleophile (to produce transient intermediates), implying that zinc(II)-bound alkoxides are more favorable than zinc(II)-OH$^-$ for overriding the highest reaction barriers in the overall processes. The chymotrypsin reaction also involves slow "acyl transfer" to the

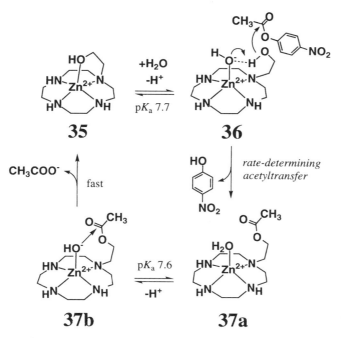

Fig. 10. An overall reaction mechanism for 4-nitrophenyl acetate hydrolysis catalyzed by an alcohol-pendant cyclen zinc(II) complex

serine OH, which is then rapidly hydrolyzed by H_2O activated by the imidazole group (Fig. 8). Less solvation around the alkoxide than around the OH⁻ may account for the stronger nucleophilicity of the alkoxide.

An alcohol in the vicinity of the zinc(II) ion can also be indirectly activated, as demonstrated by the alcohol-pendant cyclen zinc(II) complex **35** and its monodeprotonated complex **36** (see Fig. 10) [12]. The overall reaction mechanism for the catalytic hydrolysis of 4-nitrophenyl acetate was a little different from the earlier one by the [12]aneN₃ homologue **32b** (see Fig. 9). The overall second-order rate constant k_{NA} is 4.6×10^{-1} M⁻¹ s⁻¹ at 25 °C, which is ca. 10 times greater than the corresponding value of 4.7×10^{-2} M⁻¹ s⁻¹ for a reference OH⁻-bound N-methylcyclen zinc(II) complex **38**. The initial product was an "acyl-intermediate" **37** (isolated separately), as discovered with **32b**. From the pH-dependent rates and various other pieces of evidence, we concluded: (i) *The intramolecular alcohols can be bound to zinc(II) either as neutral species or as an alkoxide anion at neutral pH*. This may be an answer to the long debate on whether substrate alcohols interact with zinc(II) in ADH as neutral or alkoxide species [46]. (ii) *The zinc(II)-bound neutral alcohol may be deprotonated (or activated by zinc(II)-bound OH⁻) and act as strong nucleophiles*. The "acyl-intermediate" **37a** was then immediately hydrolyzed by the intramolecular attack of zinc(II)-bound OH⁻ in **37b**, which is generated at the vacated coordination site with a pK_a value of 7.6. For comparison of the rates, the half-life time of the

38

39

initial acyl-transfer reaction with 1 mM **36** is 26 min, while that of the second intramolecular hydrolysis reaction of **37b** is 6 s.

The zinc(II)-bound OH$^-$ in **36** serves as a base for activating the adjacent alcoholic OH, rather than as a nucleophile, because the alkoxide generated is a better nucleophile. In the subsequent attack on the "acyl-intermediate", the zinc(II)-bound OH$^-$ in **37b** becomes a nucleophile. The strong nucleophilicity of the initially neutral alcoholic OH in **36** may suggest a new reaction mechanism for zinc(II)-containing serine (or threonine) enzymes (e.g., alkaline phosphatase), whereby a fairly remote zinc(II) may possess the acidity to activate the serine OH through a strong hydrogen bond network from zinc(II)-bound OH$^-$. This model may also allow a new unprecedented proposal about the mechanism of carbonic anhydrase, in particular the function of Thr(199), which may be involved as a primary nucleophile rather than as the now accepted hydrogen bonding acceptor (see Sect. 3). A strong hydrogen-bond network with zinc(II)-bound OH$^-$ has been observed in a bis(zinc(II)-cyclen) complex stabilized by an intramolecular hydrogen bond between zinc(II)-OH$^-$ and H$_2$O-zinc(II) in **39** at neutral pH, where the two pK_a values of the zinc(II)-bound water molecules are 6.7 and 8.5, respectively, at 25 °C [47].

Recently, a benzyl alcohol-pendant cyclen zinc(II) complex **40** was synthesized. The zinc(II)-bound alkoxide species **41** generated with a pK_a value of 7.5 at 25 °C was isolated as crystals (see Fig. 11) [13]. This new zinc(II)-bound alkoxide complex reacts with bis(4-nitrophenyl) phosphate directly to yield a "phosphoryl-transfer" intermediate **42** which was isolated and characterized. The zinc(II)-bound alkoxide in **41** is a more reactive nucleophile than N-methylcyclen-zinc(II)-OH$^-$ **38**. The second order rate constant k_{BNP} of 6.6 × 10^{-4} M^{-1} s^{-1} is 125 times greater than the corresponding value of 5.2 × 10^{-4} M^{-1} s^{-1} with **38**. In anhydrous DMF solution, the initial phosphoryl transfer reaction is ca. 1,700 times faster than in aqueous solution, which is explained by poorer solvation of the nucleophile by less polar DMF. This fact is of interest in view of the hydrophobic environments at most of the active center of zinc enzymes. In the subsequent reaction of **42** the pendant phosphodiester is subject to intramolecular hydrolysis by the zinc(II)-bound OH$^-$ in **44** at a more alkaline pH to yield the phosphomonoester product **45**. The pK_a value of 9.1 for the zinc(II)-bound H$_2$O in the intermediate complex **42** was determined potentiometrically and kinetically. This value is higher than those for the previous zinc(II)-cyclen complexes (e.g., **2a** and **39**), due to the competing coordination of the pendant phosphodiester anion. The first-order rate constant for the reaction **44** → **45** is

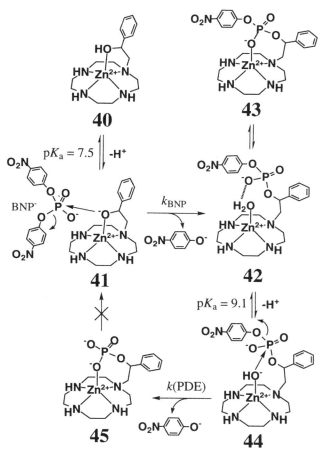

Fig. 11. Reaction mechanism for P-O ester bond cleavage of bis(4-nitrophenyl) phosphate by an alkoxide-pendant cyclen zinc(II) complex

3.5×10^{-5} s^{-1} at 35 °C. As a reference to this intramolecular phosphodiester hydrolysis, the intermolecular hydrolysis of ethyl (4-nitrophenyl) phosphodiester by N-methylcyclen-zinc(II)-OH$^-$ **39** gave a second-order rate constant of 7.9×10^{-7} M^{-1} s^{-1} at 35 °C. Thus, the intramolecular hydrolysis is about 45,000 times faster than the intermolecular hydrolysis with 1 mM **39**. These findings with the model demonstrate that the two-step reaction, involving the initial attack by the alcohols proximal to zinc(II) for the first phosphoryl transfer and the subsequent zinc(II)-OH$^-$ attack for the intramolecular phosphate hydrolysis, is kinetically more favorable than the one-step reaction of a direct intermolecular attack of zinc(II)-OH$^-$ on phosphate esters. The principle discovered by these models may be of relevance to alkaline phosphatase activity (see Fig. 7). However, unlike the situation with the enzyme, the intramolecular phosphomonoester in **45** was not hydrolyzed, because the pendant dianion strongly

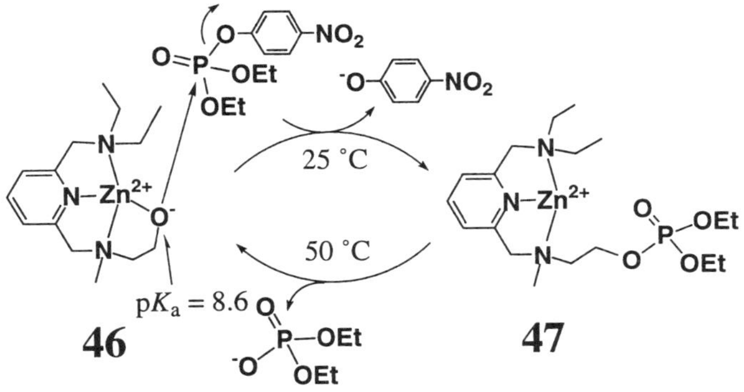

Fig. 12. Reaction mechanism for the hydrolysis of diethyl(4-nitrophenyl) phosphate by an alkoxide-pendant triamine zinc(II) complex

coordinates to zinc(II), preventing formation of the nucleophilic zinc(II)-bound OH$^-$. In alkaline phosphatase (Fig. 7), another zinc(II) (at the M1 site) can generate zinc(II)-bound OH$^-$ to attack the phosphoseryl intermediate.

Another example of the zinc(II)-alkoxide reactivity was shown by a zinc(II)-polyamine complex with an alkoxide-pendant **46** (pK_a = 8.6 for the ligand alcohol), which promotes the hydrolysis of a phosphotriester, diethyl(4-nitrophenyl) phosphate via the phosphoryl-intermediate **47** to diethyl phosphate (see Fig. 12) [48]. The higher nucleophilicity of the zinc(II)-bound alkoxide complex **46** relative to a reference zinc(II)-OH$^-$ complex **48** was evident by the hydrolysis rate at 25 °C and pH 8.6 which was 30,000 times higher. When the phosphoryl-transferred product **47** was allowed to stand at 50 °C, hydrolysis occurred to yield diethyl phosphate and regenerate the initial compound **46**. However, the kinetics and the mechanism of the second reaction were not reported.

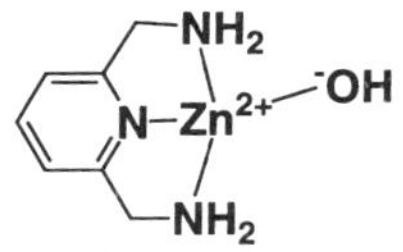

48

2.6
Activation of Alcohols with Other Metal Ions

A lanthanide(III) complex of a cyclen attached to four hydroxyethyl groups **49** promoted an attack of a pendant alcoholic OH group at a phosphodiester substrate bis(4-nitrophenyl) phosphate (BNP$^-$) (see Fig. 13) [49]. Treatment of BNP$^-$ with the europium(III) complex at 37 °C resulted in the rapid production of 4-nitrophenolate and the phosphodiester-pendant complex **51**. The phosphoryl-transfer reaction at pH 9 and 37 °C followed a second-order dependence

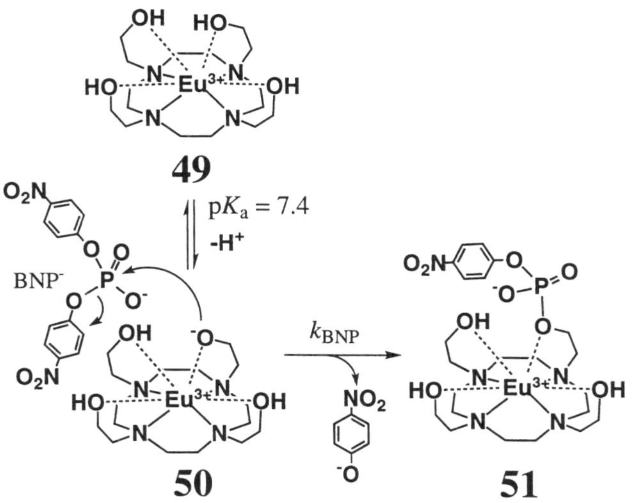

Fig. 13. Reaction mechanism for P-O ester bond cleavage of bis(4-nitrophenyl) phosphate by an alkoxide-pendant cyclen europium(III) complex

($k_{BNP} = 0.49$ M^{-1} s^{-1}) on [BNP$^-$] and the [europium(III) complex]. The pK_a of 7.4 obtained from the kinetic data at various pHs matched closely the value of 7.5 determined by potentiometric pH titration, suggesting that the active form of the europium(III) catalyst has a bound hydroxide or alkoxide group (see 50 in Fig. 13). Accordingly, the reaction mechanism with the alkoxide complex is proposed as depicted in Fig. 13. However, unlike the earlier zinc(II) systems (e.g., 41 and 46), the product phosphodiester 51 did not seem to be hydrolyzed further.

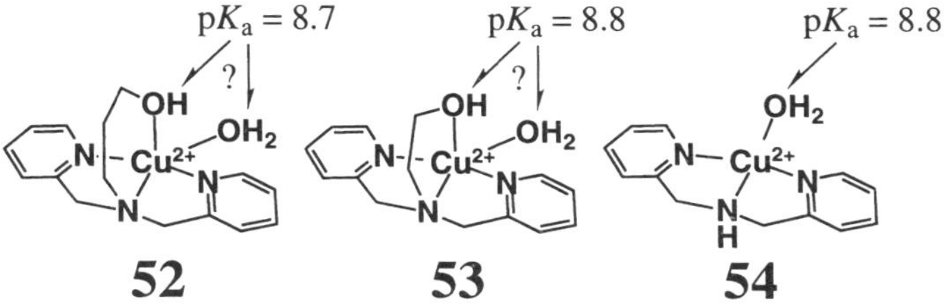

The activation of alcohol was also reported with the copper(II) ion [50]. The reactivity of alcohol-pendant copper(II) complexes 52, 53, and 54 for cleaving bis(2,4-dinitrophenyl) phosphate (BDP$^-$) was compared at various pHs. The copper(II) complex 52 with the hydroxypropyl group cleaved the phosphate by transesterification to the phosphodiester-pendant copper(II) complex 55, while the hydroxyethyl-pendant complex 53 and the pendant-less complex 54 cleaved BDP$^-$ to the phosphomonoester product by hydrolysis. It was not reported whether or not the anticipated subsequent intramolecular copper(II)-bound OH$^-$ attacked to yield a phosphate-pendant complex 56, as was found with a similar zinc(II) complex 44 [13]. The hydroxypropyl-pendant copper(II) complex **52** was much more reactive towards BDP$^-$ ($k = 0.72$ M^{-1} s^{-1} at 25 °C and pH 8.8) than **53** ($k = 9.5 \times 10^{-3}$ M^{-1} s^{-1}) and **54** ($k = 2.0 \times 10^{-2}$ M^{-1} s^{-1}). The potentiometric pH titrations of **52**, **53**, and **54** gave one titratable proton each with pK_a values of 8.7, 8.8, and 8.8, respectively. It was not clear whether the pK_a values for **52** and **53** are due to the copper(II)-bound alcohols or copper(II)-bound H$_2$O. The copper(II)-bound H$_2$O at the equatorial position should be more acidic than the apical donors. The pH rate profile gave kinetic pK_a values of 8.8, 8.9, and 9.0, respectively, which agree with those obtained potentiometrically. In the cleavage of BDP$^-$ by **52**, the first coordination of BDP$^-$ to replace the equatorial H$_2$O was followed by intramolecular copper(II)-bound alkoxide attack on the active phosphodiester. The authors concluded that it is the apical alkoxide species, rather than the thermodynamically more probable equatorial hydroxide species, which is the active form for the transesterification of **52** (see Fig. 11 for alkoxide-pendant zinc(II) cyclen system). However, the alternative, more likely mechanism of the neutral pendant alcohol activation by an intramolecular copper(II)-bound OH$^-$, was not discussed (see Fig. 10 for alcohol-pendant zinc(II) cyclen system).

55   **56**

3
Recent Progress in Protein Engineering of Carbonic Anhydrase and Alkaline Phosphatase and Future Modeling

The His$_3$-zinc(II) motif of HCA II (see Fig. 14) was converted to His$_2$Asp-zinc(II) by mutating His(94) → Asp, where the protein structure did not significantly change [51]. The effect of this replacement was seen in the zinc(II) dissociation constant K_d of 15 nM at pH 7, a 10^4-fold increase relative to the K_d of 4 pM for the wild-type HCA II. The pK_a value of the zinc(II)-OH$_2$ increased from 6.8 (wild-type) to ≥ 9.6, as determined by the pH-dependent hydrolysis of 4-nitrophenyl acetate. Its second-order rate constant decreased ca. 7-fold from 2500 (wild-type) to ≥ 365 M^{-1} s^{-1}. The large decreases in activity for CO$_2$ hydration reflect both the decreased catalytic efficiency and the increased pK_a, which, the authors concluded, arose from a weakened hydrogen bond between Thr(199) and zinc(II)-OH$_2$ in the mutant HCA II, in addition to the anionic ligand (COO$^-$) effect.

The functional importance of a conserved hydrophobic face in human carbonic anhydrase II (HCA II), including amino acid residues 190–210, was in-

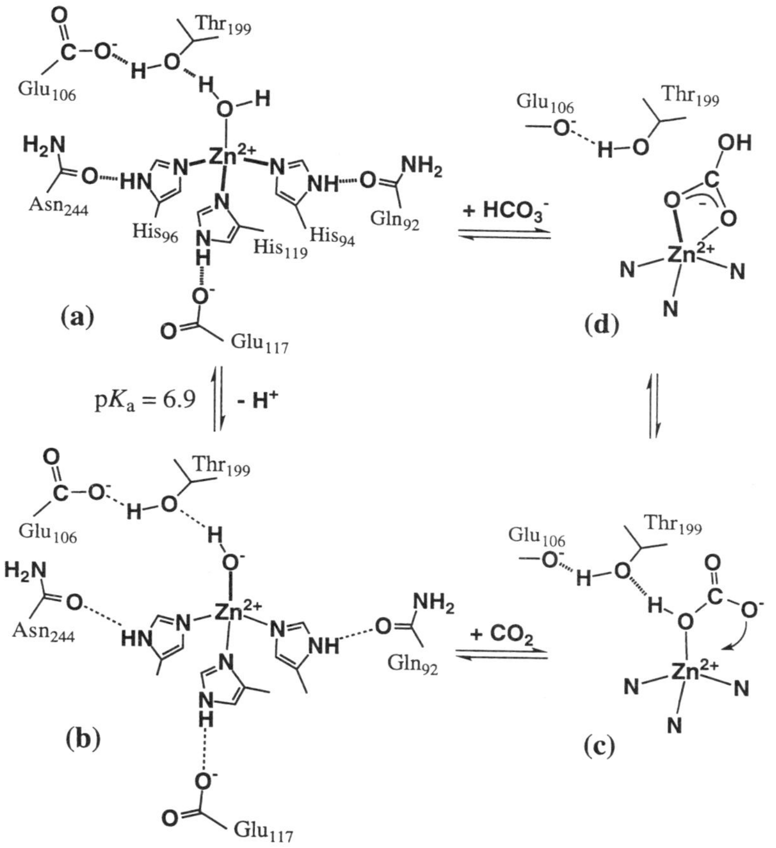

Fig. 14. Proposed mechanism of CO_2 hydration at the HCA II active center: (a) zinc(II)-OH_2 form, (b) zinc(II)-OH^- form, (c) HCO_3^--bound intermediate (transition state) involving a hydrogen bond with Thr(199), and (d) HCO_3^--bound intermediate involving no hydrogen bond with Thr(199)

vestigated by random mutagenesis [52, 53]. Amino acid substitution of Thr(199) ($\rightarrow$ Ser, Ala, Val, and Pro), especially those lacking a hydroxyl group in this side-chain, decreases the catalytic efficiency of HCA II by more than 50-fold, implicating Thr(199) as the single crucial catalytic moiety. The X-ray crystal structure indicates that the hydroxyl OH group of Thr(199) is an integral part of a conserved active site hydrogen bond network: it is a hydrogen bond donor to Glu(106) and a hydrogen bond acceptor with the zinc(II)-bound H_2O (or OH^-) (see Figs. 14a and b). The data suggest that correct formation of this hydrogen bond network is necessary for the facile nucleophilic attack of zinc(II)-bound OH^- on CO_2 or ester substrates. The structural feature may both orient and activate the zinc(II)-OH^- in the ground state and stabilise the transition state of CO_2 hydration. The side-chain hydroxyl group stabilises zinc(II)-

bound OH$^-$ relative to zinc(II)-bound H$_2$O (pK_a of 6.9 for the wild type, 7.3 for Ser, 8.3 for Ala, 8.7 for Val, 9.2 for Pro), stablises the transition state for bicarbonate dehydration (Fig. 14 c), and destabilises the HCA II-HCO$_3^-$ complex (see Fig. 14 d). An inverse correlation was discovered between logarithmic values of the second-order rate constant for CO$_2$ hydration and the pK_a values of zinc(II)-bound H$_2$O. This is a very unusual result. Brönsted correlations with positive slopes are observed for most nucleophilic reactions. The observed slope of –1 for CO$_2$ hydration indicates that interactions which destabilise zinc(II)-OH$^-$ relative to zinc(II)-OH$_2$ also destabilise the transition state for CO$_2$ hydration relative to zinc(II)-OH$^-$. The hydrogen bond network stabilises the chemical transition state and zinc(II)-OH$^-$ in a similar fashion. These data are consistent with the hydroxyl group of Thr(199) forming a hydrogen bond with the transition state and a non-hydrogen bond with the HCA II-HCO$_3^-$ complex. However, as we draw up a new proposal on the basis of our model study in Sect. 2.5, we might also postulate that the hydroxyl group of Thr(199) linked with zinc(II)-OH$^-$is an active nucleophile in the wild-type enzyme, while zinc(II)-OH$^-$ is an active nucleophile in the other mutant CA.

The functional role of the hydrogen bonds with the zinc(II)-ligands His(94), His(96), and His(119) in zinc(II) binding, the pK_a value for zinc(II)-OH$_2$, and catalysis (see Fig. 14 a), were all investigated by mutating Gln(92), Glu(117), and Thr(199) [54]. The results showed that the zinc(II) affinity is increased by a factor of 10 per hydrogen bond. The observed decrease in zinc(II) affinity is additive for the Ala(92)/Ala(117) variant, suggesting a maximal decrease of 10^4-fold for removal of all four hydrogen bonds. Thus, zinc(II) dissociation constants K_D (pM) are 4.0 for the wild-type, 18 for Ala(92), 40 for Ala(117), and 160 for Ala(92)/Ala(117). Moreover, the pK_a value of the zinc(II)-bound H$_2$O is fine-tuned when indirect ligands are altered (6.8 for the wild-type, 6.4 Leu(92), and 7.7 for Glu(92)). A plot of the logarithmic values of the pH-independent second-order rate constant for CO$_2$ hydration vs the pK_a value of the zinc(II)-bound H$_2$O for HCA II variants shows a roughly negative correlation. The slope of – 0.65 indicates that the interactions which destabilise the zinc(II)-OH$^-$ species relative to the zinc(II)-OH$_2$ (i.e., higher pK_a value) also destabilise the negatively charged pentacoordinate transition state for CO$_2$ hydration relative to zinc(II)-OH$^-$. These results suggest that the negative charge on oxygen in the transition state remains characteristic of the zinc(II)-OH$^-$ and the stabilisation of this negative charge is the determinant factor in catalysis, rather than the nucleophilicity of the zinc(II)-OH$^-$.

A similar modification of E. coli alkaline phosphatase by site-specific mutagenesis disclosed the importance of residues that are linked by hydrogen bonds to zinc(II) ligands [55]. In order to investigate the (indirect) role of the side-chain of His(372), which forms a hydrogen bond with Asp(327) (see Fig. 7), in zinc(II) binding and the catalytic phosphatase activity, His(372) was converted to Ala(372). The Ala(372) enzyme has a similar zinc(II) binding affinity as the wild-type enzyme. The hydrolytic activity (of 4-nitrophenyl phosphate) at pH 8 is 10 times lower. The mutation selectively enhances the transphosphorylation of the enzyme in the presence of a phosphate acceptor. A change in the rate-determining step at pH 8 was also observed, by which the breaking of the covalent phosphoserine bond became the rate-limiting step. These kinetic results suggest that zinc(II)-OH$^-$ becomes a weaker nucleophile in the Ala(372) enzyme. The hydrogen bond between His(372) and Asp(327) is important for neutralising the negative COO$^-$ charge on Asp(327), and therefore makes zinc(II)-OH$^-$ more stable (e.g., lower pK_a value). In the Ala(372) enzyme, Asp(327) is a stronger electron donor for zinc(II), due to the lack of the hydrogen bonding. This would raise the pK_a of the zinc(II)-bound water or make it a less effective nucleophile for the intramolecular phosphoseryl intermediate. Further evidence for this conclusion is that the Asp(327) → Asn (no charge) enzyme has higher hydrolytic activity at pH 8 [56]. We believe that this kind of fine-tuning of the zinc(II)-ligand character will gain extra recognition in future research.

Therefore, consideration of the secondary metal ligands will be of great importance for future model designs.

4
References

1. Kimura E, Koike T (1996) In: Comprehensive Supramolecular Chemistry, Vol 10:429, Pergamon
2. Kimura E (1994) Prog Inorg Chem 41:443
3. Kimura E, Shionoya M (1994) in Transition Metals in Supramolecular Chemistry, 245, John Wiley
4. Kimura E, Koike T (1991) Comments Inorg Chem 11:285
5. Kimura E, Shiota T, Koike T, Shiro M, Kodama M (1990) J Am Chem Soc 112:5805
6. Koike T, Kimura E (1991) J Am Chem Soc 113:8935
7. Zhang X, von Eldic R, Koike T, Kimura E (1993) Inorg Chem 32:5749
8. Kimura E, Koike T, Shionoya M, Shiro M (1992) Chem Lett 787
9. Koike T, Kimura E, Nakamura I, Hashimoto Y, Shiro M (1992) J Am Chem Soc 114:7338
10. Kimura E, Nakamura I, Koike T, Shionoya M, Kodama Y, Ikeda T, Shiro M (1994) J Am Chem Soc 116:4764
11. Koike T, Takamura M, Kimura E (1994) J Am Chem Soc 116:8443
12. Koike T, Kajitani S, Nakamura I, Kimura E, Shiro M (1995) J Am Chem Soc 117:1210
13. Kimura E, Kodama Y, Koike T, Shiro M (1995) J Am Chem Soc 117:8304
14. Zhang X, von Eldik R (1995) Inorg Chem 34:5606
15. Read RJ, James MN (1981) J Am Chem Soc 103:6947
16. Brown RS, Curtis NJ, Huguet J (1981) J Am Chem Soc 103:6953
17. Brown RS, Salmon D, Curtis NJ, Kusuma S (1982) J Am Chem Soc 104:3188
18. Slebocka-Tilk H, Cocho JL, Frakman Z, Brown RS (1984) J Am Chem Soc 106:2421
19. Alsfasser R, Trofimenko S, Looney A, Parkin G, Vahrenkamp H (1991) Inorg Chem 30:4098
20. Looney A, Han R, McNeill K, Parkin G (1993) J Am Chem Soc 115:4690
21. Looney A, Parkin G (1994) Inorg Chem 33:1234
22. Kitajima N, Hikichi S, Tanaka M, Moro-oka Y (1993) J Am Chem Soc 115:5496
23. Woolley P (1975) Nature, 258:677
24. Botrè F, Gros G, Storey BT (1991) Carbonic anhydrase. VCH, Weinheim
25. Khalifah RG (1971) J Biol Chem 246:2561
26. Nair SK, Calderone TL, Christianson DW, Fierker CA (1991) J Biol Chem 266:17320
27. Palmer DA, von Erdic R (1983) Chem Rev 83:651
28. Mahal G, von Erdic R (1985) Inorg Chem 24:4165
29. Ruf M, Weis K, Vahrenkamp H (1994) J Chem Soc Chem Commun 135
30. Murthy NN, Karlin KD (1993) J Chem Soc Chem Comm 1236
31. Kimblin C, Allen WE, Parkin G (1995) J Chem Soc Chem Commun 1813

32. Eriksson AE, Kylsten P, Jones TA, Lilijas A (1988) Proteins, 4:283
33. Chen RC, Kernohan JC(1967) J Biol Chem 242:5813
34. Ferrer S, Borrás J, Miratvilles C, Fuertes A (1989) Inorg Chem 28:160
35. Koike T, Watanabe T, Kimura E, Shiro M (in press) J Am Chem Soc
36. Taylor PW, King RW, Burgen ASV (1970) Biochemistry, 9:2638
37. Eklund H, Jones A, Schneider G (1986) Active site in alcohol dehydrogenase. In: Bertini I,
    Luchinat C, Maret W, Zeppezauer M (eds) Zinc enzymes. Birkhäuser, Boston p 377
38. Pettersson G (1986) Ionization properties of zinc-bound ligand alcohol dehydrogenase.
    ibid, p 451
39. Kimura E, Shionoya M, Hoshino A, Ikeda T, Yamada Y (1992) J Am Chem Soc 114:10134
40. Shoner SC, Humphreys KJ, Kovacs JA (1995) Inorg Chem 34:5933
41. Kim EE, Wyckoff HW (1991) J Mol Biol 218:449
42. Sigman DS, Jorgensen CT (1972) J Am Chem Soc 94:1724
43. Dugas H (1989) Bioorganic chemistry. Springer, Berlin Heiderberg New York, p 196
44. Cruickshank P, Sheehan JC (1963) J Am Chem Soc 86:2070
45. Vandersteen AM, Janda KD (1996) J Am Chem Soc 118:8787
46. Zeppezauer M (1986) The metal environment of alcohol dehydrogenase: aspects of
    chemical speciation and catalytic efficiency in biological catalyst. In: Bertini I, Luchinat
    C, Maret W, Zeppezauer M (eds) Zinc enzymes. Birkhäuser, Boston, p 416
47. Fujioka H, Koike T, Yamada N, Kimura E (1996) Heterocycles, 42:775
48. Kady IO, Tan B, Ho Z, Scarborough T (1995) J Chem Soc Chem Commun 1137
49. Morrow JR, Aures K, Epstein D (1995) J Chem Soc Chem Commun 2431
50. Young MJ, Wahnon D, Hynes RC, Chin J (1995) J Chem Soc Chem Commun 9441
51. Kiefer LL, Ippolito JA, Fierke CA, Christianson DW (1993) J Am Chem Soc 115:12581
52. Krebs JK, Fierke CA (1993) J Biol Chem 268:948
53. Krebs JK, Ippolito JA, Christianson DW, Fierke CA (1993) J Biol Chem 268:27458
54. Kiefer LL, Paterno SA, Fierke CA (1995) J Am Chem Soc 117:6831
55. Xu X, Qin X, Kantrowitz ER (1994) Biochemistry, 33:2279
56. Xu X, Kantrowitz ER (1992) J Biol Chem 267:16244

Zinc Catalysis in Metalloproteases

David S. Auld

Center for Biochemical and Biophysical Sciences and Medicine, Harvard Medical School and Department of Pathology, Brigham and Women's Hospital, Boston, MA 02115
E-mail: dauld@warren.med.harvard.edu

Analyses of zinc protease sites using X-ray diffraction, NMR and X-ray absorption fine structure in combination with functional studies of these enzymes are beginning to reveal the manner in which the protein modulates the properties of the zinc to achieve specificity and catalytic efficiency. The chemistry of zinc: 1) permits a variety of types and numbers of ligands and coordination geometries, 2) promotes ionization of water at neutral pH values and 3) is inert to oxido-reductants present in the medium. The type of ligands and the protein scaffolding of the zinc binding site determine the binding strength of zinc and its catalytic properties. These properties in turn are critical to the role of zinc in a wide variety of metallo-exo- and endo-protease-catalyzed processes.

Keywords: matrix metalloproteinase, aminopeptidase, carboxypeptidase, zinc enzyme, X-ray absorption fine structure (XAFS), X-ray crystallography.

Abstract 29

List of Abbreviations 30

1 Introduction 30

2 Classification of Zinc Sites in Zinc Enzymes 31

2.1 Catalytic 31
2.2 Cocatalytic 32
2.3 Structural 32

3 Influence of Protein Scaffolding on Zinc Binding and Catalytic
   Properties of Zinc 34

4 Predictive Capacity of Classifications of Zinc Metalloproteinases 36

5 Making Zinc Visible by X-Ray Absorption Fine Structure 37

6 Role of Zinc-Bound Water in ZnCPD Catalysis 37

6.1 XAFS Studies of the Free Enzyme 38
6.2 XAFS Studies of Binary and Ternary Complexes 39
6.3 XAFS and Spectroscopic Studies of CoCPD 42

7   Mechanism of Zinc Metalloproteases 43

7.1   Carboxypeptidase A 43
7.2   Matrix Metalloproteinases 44
7.3   Cocatalytic Aminopeptidases 45

8   Time-Resolved XAFS 46

9   Zinc Chemistry Pertinent to Biological Function 47

10   References 48

List of Abbreviations

NMR	nuclear magnetic resonance
XAFS	X-ray absorption fine structure
MMP	matrix metalloproteinase
LTA$_4$	leukotriene A$_4$
CPD-A or ZnCPD	zinc carboxypeptidase
CoCPD	cobalt carboxypeptidase
BLAP	bovine lens aminopeptidase
AAP	*Aeromonas proteolytica* aminopeptidase
Sar	sarcosyl
S-1, S-2, S-3	stromelysins-1, -2 and -3
Int. Col	interstitial collagenase
Macro Elas.	macrophage elastase
Gel$_{xx}$	type IV collagenases where the subscript xx refers to the MW of the protein

1
Introduction

Zinc-dependent physiological processes have been increasingly identified in the last few decades. Increasing evidence of the importance of zinc in biochemistry is a direct reflection of a number of factors: 1) the sensitivity and relative ease by which zinc can be measured with atomic absorption and emission techniques 2) availability of high-resolution structures of zinc binding sites in proteins 3) identification of families of proteins based on primary structure relationships through translation of their gene sequences. These studies have established zinc as an integral component of numerous functional proteins involved in a multiplicity of vital processes, thus accounting for its fundamental importance in metabolism, genetic transmission, growth and development [1–3].

2
Classification of Zinc Sites in Biological Systems

Structural analyses of these zinc sites by X-ray crystallography, and more recently by NMR, provide a global picture of the protein structure. X-ray absorption fine structure, XAFS, on the other hand, can probe subtle structural changes at the zinc site in a variety of physical states, including its functional state. These structures, in combination with functional studies, are beginning to give us an understanding of the interactions between the protein and zinc that has permitted zinc to orchestrate a number of chemical reactions critical to life processes. High-resolution structural studies of both zinc complex ions and enzymes now provide absolute structural standards of reference for zinc sites. There are zinc enzyme representatives in all six classes of the IUB nomenclature and high-resolution structures of members of classes I–V. About three dozen zinc enzymes are known whose zinc ligands and coordination geometries have been identified, allowing three types of zinc motifs to be recognized so far: *catalytic*, *cocatalytic* and *structural* (Fig. 1) [1, 3, 7].

2.1
Catalytic Zinc Sites

In *catalytic* sites zinc generally forms complexes with any three nitrogen, oxygen and sulfur donors of His, Glu, Asp and Cys in the binding frequency His ≫ Glu > Asp > Cys. The ligands are separated by short (1–3) and long (5–196) amino acid spacers. Water is always a ligand to the catalytic zinc. The zinc-bound water is activated for ionization, polarization or displacement by the identity and arrangement of ligands coordinated to zinc [8]. Ionization of

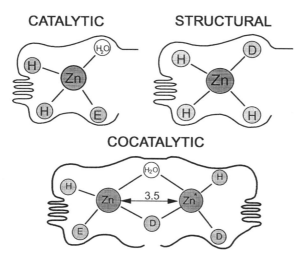

Fig. 1. Zn Sites in Enzymes: Catalytic (thermolysin [4]), Structural (catalytic domain of fibroblast collagenase [5]), Cocatalytic (*Aeromonas proteolytica* aminopeptidase [6])

the activated water or its polarization, brought about by a base from an active-site amino acid, provides hydroxide ions at neutral pH, and its displacement results in Lewis acid catalysis of the catalytic zinc atom. The structure of the active site implies that the identity of the three protein ligands, their spacing and secondary interactions with neighboring amino acids in conjunction with the vicinal properties of the active center created by protein folding are critical for the various mechanisms by which zinc can be involved in catalysis.

2.2
Cocatalytic Zinc Sites

These sites occur in multi-metal zinc enzymes where – prior to the X-ray structure determination – these metal atoms were referred to as "modulating" or "regulatory" [3]. X-ray structures of these enzymes in numbers sufficient to allow the systematization of their zinc binding sites have become available only in the course of the last few years [7]. In such sites two or three zinc atoms are in close proximity with two of the zinc sites bridged by a single amino acid residue, usually either Asp or Glu and sometimes a water molecule. While His and Cys predominate as ligands of catalytic and structural zinc atoms, Asp predominates in cocatalytic zinc sites where the frequency is Asp > His > Glu. Ser, Thr, Lys and Trp have been found to be ligands to the bridging zinc sites. The bridging Asp and H_2O ligands could have critical roles in this process. Thus, their dissociation from either or both metal atoms during catalysis could change the charge on the metal promoting its action as a Lewis acid or allowing interaction with an electronegative atom of the substrate. Alternatively, the bridging ligand might participate transiently in the reaction as a nucleophile or general acid/base catalyst. In this manner the metal atoms and their associated ligands would play specific roles in each step of the reaction that works in concert to bring about catalysis.

2.3
Structural Zinc Sites

The first *structural* zinc sites, found in alcohol dehydrogenase (ADH) and aspartate transcarbamylase, contained four Cys bound to the zinc [9, 10]. The Cys ligands in ADH and aspartate transcarbamylase are separated by 2, 2, 7 and 4, 22, 2 amino acids, respectively. There is no coordinated water molecule in these sites.

The matrix metalloproteinases, MMPs, contain two zinc atoms [11, 12], one of which has the catalytic zinc binding signature of the astacin family [13 – 16]. The second zinc site identified by structural analyses of the catalytic domains of fibroblast, MMP-1 [5, 17, 18], and neutrophil, MMP-8 [19 – 21], collagenases, stromelysin-1, MMP-3 [22 – 24] and matrilysin, MMP-7 [25], bound with synthetic inhibitors, may be the first example of a non-cysteinyl structural zinc binding site in an enzyme. It is comprised of three histidine ligands (His-168, His-183, and His-196) and an aspartate (Asp-170) tetrahedrally coordinated to the zinc with spacing intervals of 1, 12 and 12 (Fig. 2). This zinc site is unique to this sub-family of the extended astacin family [13–16]. Sequence alignment of the MMPs and other homologous members indicates that these four ligands and the spacing intervals are highly conserved (Fig. 2). Conservation of the residues adjacent to all the ligands is much higher than the overall conservation observed for this large group of homologous proteins, suggesting this is probably an important region for the structure and/or function of these proteins. This zinc site has the characteristics of a structural zinc site in that there are no bound water molecules and a relatively short sequence of the protein provides all four protein ligands [1, 3]. However, although the two zinc atoms are 12.5 Å

		168	170			183			196		
Human Cat. Domain	E	H G D F		L A	H	A Y	D A	H	F D
Rabbit S-1	E	H G D F		L A	H	A Y	D A	H	F D
Human S-2	E	H G D F		L A	H	A Y	D I	H	F D
Rat S-2	E	H G D F		L A	H	A Y	D A	H	F D
Mouse S-1	E	H G D F		L A	H	A Y	D A	H	F D
Rat Int. Col.	E	H G D F		L A	H	A Y	D A	H	F D
Bovine Int. Col.	D	H R D N		L A	H	A F	D A	H	F D
Pig Int. Col.	D	H R D N		L A	H	A F	D A	H	F D
Human Int. Col.	D	H R D N		L A	H	A F	D A	H	F D
Rabbit Int. Col.	D	H R D N		L A	H	A F	D V	H	F D
Human Neut. Col.	D	H G D N		L A	H	A F	D A	H	F D
Mouse Int. Col.	E	H G D F		L A	H	A F	D A	H	F D
Mouse Macro. Elas.	A	H G D F		L A	H	V F	D A	H	F D
Human Macro. Elas.	A	H G D F		L A	H	A F	D A	H	F D
Human S-3	W	H G D D		L A	H	A F	D V	H	F D
Frog S-3	W	H G D N		L A	H	A F	D V	H	F D
Envelysin	D	H G D G		L A	H	A F	D A	H	F D
Human Matrilysin	A	H G D S		L A	H	A F	D A	H	F D
Rat Matrilysin	D	H G D N		L G	H	A F	D A	H	F D
Cat Matrilysin	A	H G D F		L A	H	A Y	D A	H	F D
Mouse Gel-92	E	H G D G		L A	H	A F	D A	H	F D
Human Gel-92	E	H G D G		L A	H	A F	D A	H	F D
Bovine Gel	E	H G D G		L A	H	A F	D A	H	F D
Mouse Gel-72	E	H G D G		L A	H	A F	D S	H	F D
Human Gel-72	E	H G D G		L A	H	A F	D S	H	F D

Fig. 2. "Structural" zinc binding site signature of the matrilysin family of metalloproteases. Abbreviations: S-1, S-2, S-3 refer to stromelysins-1, -2 and -3. *Int. Col* is interstitial collagenase and *Macro Elas.* refers to macrophage elastase [12]. *Gel*$_{xx}$ is the type IV collagenases where the subscript xx refers to the MW of the protein. Histidine ligands are shown in *bold face*. The ligand numbers are assigned on the basis of the results of the X-ray structure of the catalytic domain of fibroblast collagenase [5]

apart, several of the conserved amino acids adjacent to the third and fourth His ligands to the second zinc site (e.g. Ala-182, -184, -195 and Phe-197) form the environment around the predicted catalytic residue Glu-218 (data from Brookhaven sources, CR Chong & DS Auld unpublished observations). The properties of the "structural" zinc site in the MMPs and in other such sites may therefore indirectly influence function by effecting the local conformation.

3
Influence of Protein Scaffolding on Zinc Binding and the Catalytic Properties of Zinc

The binding constants for imidazole and acetate binding to zinc are not particularly large, e.g. pK_1 for acetate is ~ 1.5 [26]. However, the binding constant for ethylenediaminetetraacetic acid, EDTA, is 10^{17}, probably due, in part, to the proximity of the four carboxylates. The fact that so many zinc enzymes bind zinc with picomol binding constants, yet have imidazole and acetate as their ligands, seems to reflect their proximity and restricted mobility in the protein. The zinc binding sites frequently have an α-helix or β-sheet structural region of the protein that supplies the zinc ligands, particularly in catalytic zinc sites [3]. There is a strong correlation between the short spacer length and the type of secondary support structure that supplies the ligand. Thus, in carbonic anhydrase, the secondary support structure is a β-sheet and the spacer is one. This allows the ligands to come from the same side of the sheet. In contrast, when an α-helix support structure is used, as in thermolysin, the spacer consists of three amino acids again juxtaposing the ligands in an orientation suitable for establishing a tetrahedral-like coordination sphere.

Secondary interactions of the ligands with hydrogen-bonding groups of either side-chains of the amino acid residues or the carbonyl oxygen of the backbone peptide chain may be critical to the formation and stabilization of the zinc sites containing oxygen and nitrogen ligands. Comparative structural studies of four of the first known zinc enzymes, carbonic anhydrase, carboxypeptidase A, alcohol dehydrogenase and thermolysin led to the identification of carbonyl and carboxyl "orienters" [27]. The particular interaction between a γ or δ carboxyl group of Asp or Glu, respectively, and a histidine ligand to the zinc was referred to as a new catalytic triad [28]. Such interactions occur for instance between the δ-carboxylate of Glu-117 and the His-119 ligand of carbonic anhydrase and the γ-carboxylate of Asp-170 and the His-142 ligand of thermolysin. Secondary interactions are postulated to be involved in strengthening the metal complexation or modulating the nucleophilicity of the zinc-bound water [28]. Both have been proven to occur in the case of carbonic anhydrase in a series of elegant structure-function studies on mutant carbonic anhydrases [29–31].

Carbonic anhydrase converts carbon dioxide into bicarbonate. The three His ligands of these enzymes readily allow formation of zinc hydroxide which can then add OH$^-$ to CO_2 to form HCO_3^- [32]. In carbonic anhydrase II the three protein zinc ligands, His-94, His-96 and His-119 form hydrogen bonds with the carboxamide of Gln-92, the backbone carbonyl oxygen of Asn-244 and the carboxylate oxygen of Glu-117, respectively. The functional and structural consequences of mutating these residues have recently been examined. The results indicate that the zinc affinity is reduced by about a factor of five- to ten-fold when a native H-bond is eliminated. The effects also appear to be additive. Thus the combined mutations of Gln92Ala and Glu117Ala lead to a 40-fold increase in the dissociation constant for zinc while the individual mutations lead to 4- and 10-fold increases, respectively [30]. The weakened binding is largely accounted for by an increased off-rate for zinc dissociation.

These secondary protein interactions would also be expected to effect the charge on the zinc. The binding of hydrogen-donating or hydrogen-accepting groups to the zinc ligands might therefore modulate the pK_a of the catalytic zinc-bound water resulting in changes in the nucleophilicity of the water and and account for the ease at which it can expand its coordination shell or allow the zinc-bound water or hydroxide to be displaced. In the case of the Gln92Glu mutation in carbonic anhydrase II, a neutral hydrogen bond acceptor is replaced by a negatively charged acceptor [31]. The positive charge on the zinc should therefore be reduced, which in turn should make the ionization of the water more difficult. The results of the functional studies confirm this relationship, since the kinetically determined pK_a for the water increases by 0.7 units for this mutation [30].

The importance of carrying out both structural and functional studies is also demonstrated by these studies. The pK_a values for the native and Glu117Ala mutant are 6.8 and 6.9, respectively. The value for the Ala mutant should have dropped by as much as one unit due to the loss of the anionic interaction with the His ligand. However, X-ray studies demonstrate that the Glu-117 carboxylate-His-119 interaction is replaced by a chloride ion in the case of the Glu117Ala mutant. Thus the Glu-117 hydrogen-bond to His-119 probably serves two purposes: 1) to enforce restricted mobility of the His-119 ligand leading to increased binding of the zinc and 2) to increase the nucleophilicity of the zinc by raising its pK_a. The chloride ion can replace the negative charge of the carboxylate but is probably less effective at restricting the motion of the histidine ligand.

The importance of the secondary interactions to zinc binding indicate they must be considered in any attempt to make a "synthetic" zinc enzyme. They also point out the dangers that can occur when only the functional consequences of mutations are used to make deductions. Recent mutagenesis of the β-lactamase from *Bacteriodes fraglis* led the authors to conclude that a second zinc binding site in this enzyme was composed of four Asp ligands [33]. The X-ray structure of this enzyme shows that this zinc is bound to the side chains of Asp, His and Cys residues [34]. Only one of the four postulated Asp residues directly ligates the zinc. The structure reveals that the carboxylate of one of the other predicted Asp ligands forms a hydrogen bond with a His ligand of the first zinc site while another has no direct contact but probably aids in positioning a Trp residue near the three His ligands at the first zinc site. The mutagenesis studies therefore revealed the importance of these residues for the structure and or function of the enzyme but led to the wrong conclusion in identifying the ligands linked directly to the zinc.

4
Predictive Capacity of Classifications of Zinc Metalloproteinases

In the case of metalloproteases there have been several recent attempts at classifying them. These have been based on primary sequence relationships [1, 3, 13–16, 35, 36], the number and types of zinc sites [1, 3], their 3-dimensional structural characteristics [15, 16] and evolutionary concerns [36]. These classifications have been used for predictions of new zinc enzymes and even for a new function of an existent enzyme. They have been used to predict catalytic zinc binding sites of enzymes whose structures have not yet been determined, e.g. mono-zinc aminopeptidases and leukotriene A_4, LTA_4 hydrolase [1]. In the mono-zinc human intestinal, rat kidney and *E. coli* aminopeptidases, a domain of 300 amino acids contains two His and one Glu, similar to the zinc binding site of thermolysin. The three amino acids of the short spacer in thermolysin are identical to those of the mono-zinc aminopeptidases. The long spacer in thermolysin contains 19 amino acids, corresponding to 18 amino acids in aminopeptidases [1]. The presence of an amino acid segment in leukotriene A_4 hydrolase that is homologous to the zinc binding domain of intestinal aminopeptidase indicated the presence of zinc and a hitherto unrecognized aminopeptidase activity in LTA_4 hydrolase. On this basis, the leukotriene A_4 hydrolase was shown to contain 1 g-at zinc/mol protein, to exhibit aminopeptidase activity and to be inhibited by bestatin and captopril, specific peptidase inhibitors [37, 38]. In addition, mutagenic replacements of the proposed LTA_4 hydrolase zinc ligands cause complete loss of zinc and abolish its activity towards both types of substrates suggesting that these residues are the probable zinc ligands [39]. The fact that aminopeptidase inhibitors also inhibit LTA_4 hydrolase leads to new avenues for drug development based on inhibition of these enzymes.

These zinc classifications can also predict new zinc binding sites. In 1988 *Astacus* protease, or astacin as it is now called, was identified as a zinc protease [40]. In one amino acid segment the presence of two histidines, HExxH, suggested this zinc site could be similar to that of thermolysin [4]. However it was not identical since no Glu was found 19 amino acids removed from the proposed second His ligand. Two other zinc proteases, *Serratia* protease [41] and fibroblast collagenase [42, 43], in addition to *Astacus* protease [44], did have a very similar clustering of potential zinc ligands, HExxH xGxxH [40]. By 1990 a search of protein and translated gene banks with this potential zinc binding site signature had led to the retrieval of only a handful of proteins, all of which were believed to be zinc proteases [45]. These included *Serratia* protease, Protease B, the snake venom protease, Ht-d., and all the collagenases, stromelysins and gelatinases. By 1992 searching with this sequence had led to the identification of 33 proteases that now defined four major groups of homologous proteins with this zinc binding site signature [13]. The recognition of the homology of both mouse kidney and human intestine (then known as PPH hydrolase) meprin to *Astacus* protease led in fact to the naming of the immediate Astacin family of zinc proteases [46, 47]. This discovery was rapidly followed by the X-ray crystallographic structure of astacin that confirmed the prediction of this new catalytic zinc site [48]. Within two years the structures of a member of each of the Astacin subfamilies was determined, i.e. matrix metallo-proteinase-1 [6], the snake venom protease, adamalysin II [49] and the alkaline protease of *Pseudomonas aeruginosa* [50]. Remarkably these proteases not only have the same catalytic zinc binding site but are also topologically "homolo-gous", even though they would not have been considered homologous based on their primary sequences. The zinc binding site in this superfamily is the smal-lest site known since all zinc ligands (His-218, -222 and -228) and the presumed catalytic residue, Glu-219, are supplied from an eleven amino acid segment. However the first two His ligands form hydrogen bonds with the backbone amides of residues 216 and 235 indicating that the protein scaffolding is still important.

5
Making Zinc Visible by X-ray Absorption Fine Structure

Direct examination of the role of zinc in biological systems has not been pos-sible since zinc has a full d-shell and thus no useful chromophoric properties, like cobalt, copper or iron, to reveal its presence. However, X-ray absorption fine structure, XAFS, experiments do not need a metal with an unfilled d-shell in or-der to examine the spectral properties of the metal in different environments [51]. Our initial XAFS studies have focused on the solution form of ZnCPD, examining the effect of H^+ ion, substrate and inhibitor binding on the coordi-nation properties of the catalytic zinc site [52–55]. These studies have revealed the identity of the group responsible for the alkaline pK_a in activity profiles and determined the structure of intermediates during catalysis. The results indicate that the zinc-bound water plays a central role in all of these processes.

6
Role of the Zinc-bound Water in ZnCPD Catalysis

Analysis of the individual k_{cat} and K_m profiles for ZnCPD are supportive of a 3 protonation state model [56–58] (Fig. 3). According to this scheme, when the enzyme is in its EH_2 form below the acidic pK_{EH2} it can bind substrates but not hydrolyze them. The ionization of the group EH_2 with a pK_a of 6.0 transforms the enzyme into its active form EH. Further ionization of the enzyme to the E form, occurring with a pK_{EH} of 9.0, markedly reduces substrate binding and therefore catalysis.

Several reasons have led to the assignment of pK_{EH2} to the ionization of the carboxyl group of Glu-270 and its subsequent interaction with the water ligand of the zinc, stabilizing the active site structure ($EH_2 \Leftrightarrow EH$, Fig. 3). This is con-sistent with the crystallographic studies which show that the interatomic distance between the zinc and the carboxylate is 4.5 Å, a distance consistent with an H-bond between the metal-bound water molecule and Glu-270 [59]. Chemical modification of Glu-270 with CMC inactivates the enzyme [60]. Temperature-jump studies provide evidence for a conformational change coup-led to the formation of the EH_2 form of the enzyme which might reflect move-ment of Glu-270 away from the metal [61]. Spectrokinetic and NMR studies on

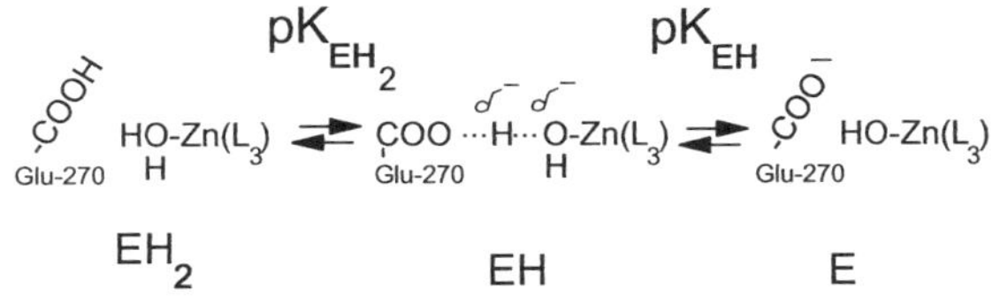

Fig. 3. Ionizable groups critical to ZnCPD catalysis

CoCPD show that anions bind to the EH_2 form of the enzyme. This is reasonable if the protonation of Glu-270 breaks its interaction with the water and thus allows its displacement by anions [62–65] (Fig. 3). On the other hand the enzyme is inhibited by zinc hydroxide by binding to the EH form of the enzyme [66, 67]. This is also consistent with this model since the positive $ZnOH^+$ can bind to the negative Glu carboxylate oxygen and displace the metal-bound water forming a Zn-O-Zn bridge. X-rax diffraction studies have confirmed this mode of Zn inhibition [92].

6.1
XAFS Studies of Free Enzyme

Recent XAFS and spectroscopic studies have aided in identifying the group probably responsible for the alkaline pK_a. Examination of the XAFS spectrum of the zinc enzyme over the pH range 7–11 reveals a near edge feature that titrates with pH (Fig. 4). The intensities of the double peak feature located at 9659 and 9664 eV at pH 7.2 invert at pH 10.6 (pH values at freezing point of solution, $-4\,°C$). Data analysis yields two distributions of atoms in the first coordination shell of ZnCPD at all pH values and at both 150 K and 297 K [52]. Direct comparison of the first and higher coordination shells of ZnCPD reveals structural differences between pH 7.2 and 10.6 that are principally reflected in the average inner coordination distance decreasing from 2.024 to 2.002 Å (Table 1). The structural analyses of ZnCPD at intermediate pH values show this distance decreases as a function of the increase in pH [52]. The XAFS Debye-Waller factor shows an increased structural disorder for the four atom distribution at the alkaline pH value, while the higher shell comparison shows that the ligands His-69 and His-196 remain unchanged as the pH is increased. The plot of the normalized spectral differences between the peaks located at 9659 and 9664 eV of the absorption edge as a function of pH conforms to a theoretical pH-titration curve with a pK_a of 9.49 at $-4\,°C$ (52, Fig. 5). This value corresponds to that obtained by extrapolation of the temperature-dependent value of pK_{EH} obtained by kinetic analysis of the CPD-A-catalyzed hydrolysis of tripeptides [57]. The pH-dependent shorter average metal-ligand distance and the increased first shell structural disorder is most easily explained by one of the two metal ligands (the water or the ε_1 oxygen of Glu-72) moving 0.09 ± 0.03 Å closer to the zinc ion at alkaline pH values. The structural changes can be attributed to ionization of the water coordinated to the catalytic zinc ion or an ionization-linked alteration in the position of Glu-72 [52]. Since the kinetics indicate that this ionization leads to weaker substrate binding, it is reasonable to

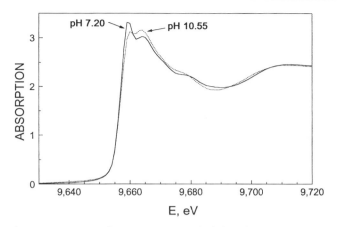

Fig. 4. Zn K-edge XAFS spectra of ZnCPD at pH 7.2 (*solid*) and 10.6 (*circles*) at –4 °C [52]

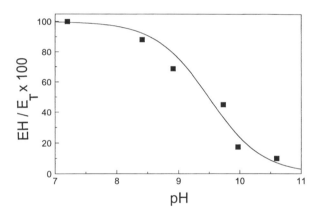

Fig. 5. pH dependence of the spectral difference of peaks at 9.659 & 9.664 KeV for [52]

assign pK_{EH} in the kinetic profiles to the ionization of the metal-bound water (EH ⇔ E, Fig. 3). This ionization strengthens the Zn-OH bond and prevents displacement of hydroxide by substrates and inhibitors.

6.2
XAFS Studies of Binary and Ternary Complexes

Examination of the effect of pH on the XAFS spectra of binary and ternary product complexes of ZnCPD provides further evidence for the role of unionized water in catalysis [54]. Inhibition studies show that L-Phe, the amino acid product of the most rapidly turned over peptide substrates of CPD-A, shifts pK_{EH} from 9 to 7.5 at 25 °C [58]. Increasing the pH from 7 to 10 for the ZnCPD · L-Phe complex results in the same type of progressive spectral changes in the near-

Table 1. Structural Parameters for the Catalytic Metal of Zn- and CoCPD and their Inhibitor[a]

Enzyme	N[b]	R[c], Å	σ^2,[d] Å^2	N[b]	R[c], Å	σ^2,[d] Å^2	Q[e]
ZnCPD pH 7.0 (7.2)	4	2.024 (6)	– 0.0006	1.3 (4)	2.54 (5)	– 0.0012	1.0
ZnCPD · L-Phe pH 6.56 (6.94)	4	2.021 (6)	0.000	1.4 (5)	2.54 (5)	– 0.002	1.0
ZnCPD · L-Phe · N$_3^-$ pH 7.0 (7.2)	3.9 (5)	1.995 (6)	0.000				0.7
ZnCPD · L-Phe · N$_3^-$ pH 8.81 (9.36)	4.0 (5)	1.995 (6)	0.000				0.6
CoCPD pH 7.0 (7.2)	4	2.08 (1)	0.002	1.0	2.50 (4)	0.000	0.6
CoCPD · L-Phe · N$_3^-$ pH 7.0 (7.2)	4.1 (5)	2.046 (7)	0.001				0.6

[a] First shell fitting results using the least-squares method of analysis [54]. The pH values given in parentheses are those of the solution at its freezing point, – 4 °C.
[b] Zinc coordination number.
[c] Average absorber-scatterer bond lengths.
[d] Mean-square deviation in absorber-scatterer in Angstroms squared.
[e] Fitting criteria.

edge XAFS spectrum as are seen for ZnCPD, but the changes are complete more than 1 pH value below that observed for ZnCPD [54]. The XAFS results show that the average interatomic distance, R, for the zinc ligands of the EH · L-Phe complex decreases by 0.02 Å upon formation of the E · L-Phe complex, essentially identical to that obtained for the EH and E forms of the native enzyme (Table 1, [51]).

The results give a structural basis for the kinetic studies which show that E binds the protonated form of L-Phe more tightly than EH, thus in effect decreasing the value of pK_{EH} [58]. The protonated amino group of L-Phe probably forms a salt bridge with the carboxylate of Glu-270 thus breaking up the hydrogen bond between the carboxylate and the metal-bound water (Fig. 6). While no X-ray crystallographic results have been reported for the L-Phe complex, they have for the D-Phe complex [68]. In this case the α-amino group of D-Phe hydrogen bonds to the carboxylate of Glu-270 which no longer interacts with the metal-bound water. Similar interactions for L-Phe might therefore be anticipated. Since the water is now no longer H-bonded it is free to ionize at a lower pH value. The apparent pK_a of the zinc-bound water in this inhibitor complex is about 1 unit higher than observed in carbonic anhydrase II [30–32]. The higher pK_a value for ZnCPD probably reflects the replacement of a His ligand in carbonic anhydrase by a Glu ligand in carboxypeptidase. The presence of the Glu ligand should reduce the charge on the zinc and thus make it more difficult for the zinc-bound water to ionize.

Addition of azide to the ZnCPD · L-Phe complex at pH 7 markedly changes the zinc coordination sphere from 4 N/O atoms at 2.021 ± 0.006 Å and 1.4 ± 0.5

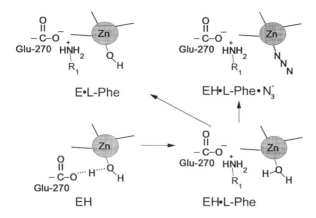

Fig. 6. Different active site forms of ZnCPD

N/O atoms at 2.54 ± 0.5 Å to 3.9 N/O atoms at 1.995 ± 0.006 Å. Examination of the higher coordination shell between 2.8 and 4.0 Å reveals that marked changes occur when azide is bound, providing further support for this assignment [54]. This is the region where the second and third nitrogens of the zinc-bound azide would be expected to reside. Both single and multiple scattering paths of these nitrogens can lead to a focusing effect that can account for the observed spectral changes. The decrease in R of about 0.03 Å in both the Zn- and CoCPD · L-Phe · N_3^- complexes is probably due to the ligand exchange from a neutral water to an anion (Table 1). The XAFS spectra of the ternary complex is pH-independent from pH 7 to 9 (Fig. 7) in agreement with the ionization of the water being the source of the spectral changes in the free enzyme and its binary L-Phe complex. The enzyme · azide · L-Phe complex is probably bound in a manner analogous to that expected for a post-transition state in a bi-product

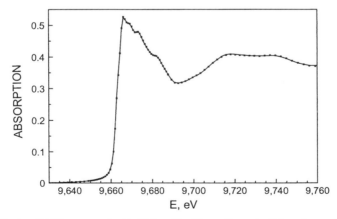

Fig. 7. Zn K-edge XAFS spectra for ZnCPD L-Phe N at pH 7.0 (*solid*) and 8.81 (*squares*) at 25 °C [54]

complex for peptide hydrolysis, i.e. the carboxylate anion of the peptide bound to the Zn and the protonated form of L-Phe H-bound to the catalytic Glu-270 carboxylate (Figures 6 & 7).

6.3
Spectroscopic Studies of CoCPD

The results of electronic absorption studies of CoCPD also suggest that azide binds to the cobalt atom. The electronic absorption spectrum of CoCPD · L-Phe · N_3^- has a band at 308 nm with a molar extinction coefficient of 1250 $cm^{-1} M^{-1}$ characteristic of a charge transfer complex and a new λ_{max} at 590 nm with a markedly elevated extinction coefficient, 330 $cm^{-1} M^{-1}$ (Table 2, [63]). The XAFS studies show that the fifth metal ligand in the free enzyme and its L-Phe complex at 2.5 Å have been farther displaced in the ternary azide complexes of the zinc and cobalt enzymes. This change in coordination properties may reflect an alteration in the Glu-72 γ-carboxylate position from bidentate to monodentate ligation. ^1H NMR spectroscopy of the isotropically shifted signals in CoCPD, CoCPD · L-Phe and CoCPD · L-Phe · N_3^- are also consistent with the XAFS observations [64, 65]. The isotropically shifted signals a, c and d of CoCPD were assigned to the NH-proton of His-69 and to the C-4 protons of His-69 and His-196, respectively. These assignments were made on the basis of the effect of D_2O on the signals and the fact that the hydrogen bond between Asp-142 and His-69 probably prevents this proton from undergoing fast exchange with the solvent [28, 64]. The signals are not shifted in the CoCPD · L-Phe complex (Table 2), in agreement with the close similarity of the XAFS spectra of the corresponding zinc complexes in their back transformed 2.8 – 4 Å spectra [54]. In addition a fourth signal, b, observed in the NMR was tentatively assigned to the γ-CH_2 protons of the Glu-72 ligand [64]. This signal shifted only 2 ppm in the CoCPD · L-Phe complex but 17 ppm in the CoCPD · L-Phe · N_3^- complex (Table 2). The XAFS results on the cobalt and zinc enzymes indicate that this assignment is probably correct since the position of the second oxygen of the Glu-72 ligand is not greatly changed in the L-Phe complexes but is shifted markedly in the enzyme · L-Phe · azide complex (Table 1).

Table 2. Spectral Properties of Inhibitor Complexes of CoCPD

Enyme Complex	Electronic Absorption λ_{max}, nm (ε, $M^{-1} cm^{-1}$)		^1H NMR[a] Chemical Shift, ppm			
			a	c	d	b
CoCPD	555 (150)	572 (150)[b]	62	52	45	56
CoCPD · L-Phe	555 (200)	574 (205)[b]	62	52	45	58
CoCPD · L-Phe · N_3^-	308 (1250)	590 (330)[c]	64	55	43	73

[a] isotopically shifted signals from protons of residues ligated to the cobalt [64].
[b] data of [69].
[c] data from [63].

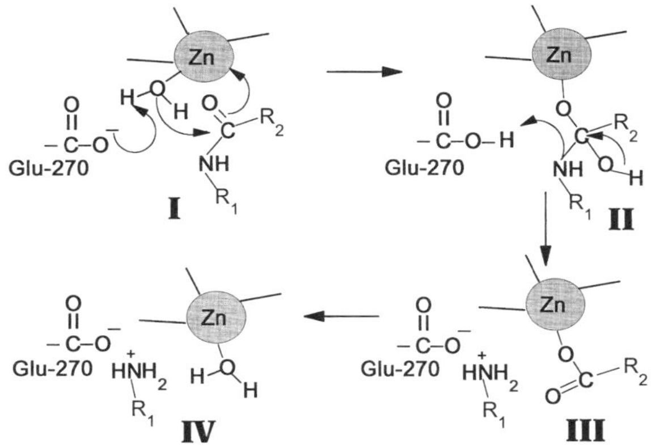

Fig. 8. Mechanism of Carboxypeptidase A catalysis

The zinc-bound water therefore affects both the acid and alkaline pK_a values of carboxypeptidase A. It needs to be in its protonated state ($[ZnL_3(H_2O)]^+$) to be catalytically active. In this way it can prepare the ionized carboxylate Glu-270 for catalysis (Fig. 8). Ionization of the water decreases the charge on the zinc making it a poorer Lewis acid for interaction with the substrate.

7
Mechanism of Zinc Metalloprotease Catalysis

7.1
Carboxypeptidase A

Cryospectrokinetics of zinc and cobalt carboxypeptidase reveal two intermediates in the hydrolysis of peptides and furnish all the rate and equilibrium constants for the reaction scheme $E + S \Leftrightarrow ES_1 \Leftrightarrow ES_2 \Rightarrow E + P$ [70–72]. The chemical and kinetic data indicate that neither of these is an acyl intermediate [73, 74]. Our XAFS studies of the effect of pH and inhibitors on the zinc coordination sphere [52–55] in conjunction with those of cryospectrokinetics [70–74] allow the following mechanism to be proposed for CPD-A catalysis (Fig. 8). In the first step, **I**, the zinc acts as a Lewis acid catalyst by expanding its coordination sphere to accept the peptide carbonyl. Glu-270 acts as a general base removing a proton from the metal-bound water allowing the hydroxide to attack the peptide carbonyl. In the next step, **II**, Glu-270 acts as a general acid catalyst by donating a proton to the leaving amine as the metal-bound tetrahedral intermediate collapses to products, **III**. Arg-127 probably assists the zinc as an electrophile in stabilizing the transition state in steps **I–II** [75,76]. In step **IV**, the N-terminal product leaves and water returns to the metal with the C-terminal product still bound in a salt-bridged manner to Glu-270. The enzyme complex is thus poised for the reverse reaction, synthesis of a peptide bond. In this state the zinc-bound water is "vulnerable" to the displacement of water by anions. This configuration of ligands, activated water and Glu-270 allows the donation of the 2 protons needed to the amino acid leaving group. This would be an example of polarization-assisted catalysis [8]. The presence of the Glu ligand reduces the charge on the zinc, retarding the ionization of the zinc-bound water.

7.2
Matrix Metalloproteinases

It is largely believed that this mechanism applies not only to carboxypeptidase [76, 77] but also to thermolysin [4] and the matrix metalloproteinase family [25] as well. However, our recent pH studies of matrilysin catalysis of Dns-PLA↓LWAR (↓ denotes bond cleavage) indicate that the pK_a values for the protein groups controlling catalysis of peptides are 4.3 and 9.6 [78, Fig. 9). The acidic pK_a value for matrilysin, 4.3, is much lower than that for the carboxy-peptidase A-catalyzed hydrolysis of peptides, 6.0 [56]. The environment of the proposed general base catalyst, Glu-218, in matrilysin is quite hydrophobic since it is surrounded by the side-chains of the highly conserved Ala and Phe residues that border the 3rd and 4th His ligands to the "structural" zinc site (Fig. 2). The effect of these residues might be to shift the pK_a of the Glu-218 residue to alkaline pH values. It would then remain in a protonated state in the ground state of the enzyme. In addition, the catalytic zinc of these enzymes is made up of three His residues, none of which are H-bonded to an adjacent Asp residue as in carboxypeptidase and thermolysin. The absence of an inner Glu carboxylate anion ligand and indirect Asp carboxylate hydrogen bonds to the His ligands both would be expected to make the charge on the zinc greater in matrilysin than in thermolysin and carboxypeptidase. This in turn could reduce the pK_a value for the zinc-bound water. An alternative mechanism to the polarization-assisted catalysis mechanism described above for carboxypepti-

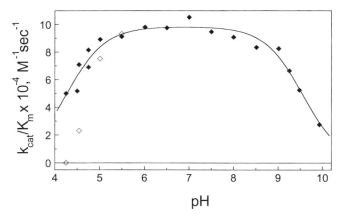

Fig. 9. Effect of pH on matrilysin activity. The assays below pH 5.5 are performed in the presence (◆) and absence (◇) of zinc[77]. The line is theoretical using an equation describing a bell shaped pH-dependence for k_{cat}/K_m and values of $pK_{a1} = 4.26$, $pK_{a2} = 9.55$ and k_{cat}/K_m of 9.8×10^4 M^{-1} s^{-1}

dase may then be envisioned for the matrix metalloproteinases. The lower pK_a for these enzymes might be the ionization of the zinc-bound water instead of the glutamate carboxylate. In this case the mechanism might involve nucleophilic attack by zinc hydroxide assisted by general acid catalysis or stabilization of the transition state by Glu-218.

7.3
Cocatalytic Aminopeptidases

The structures of two cocatalytic zinc proteases, bovine lens amino peptidase, BLAP, and *Aeromonas proteolytica* aminopeptidase, AAP, have been determined so far. The proposed catalytic mechanisms for these two enzymes are largely based on structural data on the free and inhibitor-bound enzymes supplemented by steady-state kinetic data on the effect of metals on activity. Both enzymes exhibit similar kinetics [7, 79, 80] and have C-terminal catalytic domains that have similar folds [81]. In spite of this, the two enzymes show notable differences in their cocatalytic zinc binding sites that may underlie differences in their catalytic mechanisms [81].

The hexameric leucine aminopeptidase from bovine lens was the first structure to be determined [82–86]. The zinc binding site is composed of two zincs separated by 2.91 Å [82]. Zn1 is defined as the fast-exchange, weak-binding site, where Mg can substitute for Zn [84]. It is coordinated to the carboxylate $O\gamma1$ of Asp-255, the carboxylate $O\gamma2$, the backbone carbonyl of Asp-332 and the bridging carboxylate $O\varepsilon2$ of Glu-334. The second zinc, Zn2, is defined as the tight-binding, slow-exchange site. Since the Zn2 site is absolutely essential to catalysis, this zinc has also been referred to as the catalytic site, while Zn1 is referred to as the regulatory site [87]. Zn2 is coordinated to the carboxylate $O\gamma1$ of Asp-255, the carboxylate $O\gamma1$ of Asp-273, the side chain amino N of Lys-250 and the bridging carboxylate $O\varepsilon1$ of Glu-334. No bound water molecules have been observed in the free enzyme or in any of its inhibitor complexes [86]. Nevertheless, a general base-catalyzed mechanism is still favored, using the zinc ligand Asp-255 as the general base and Lys-262 as the general acid catalyst [81, 82, 85, 86]. It has been proposed that Zn2 is involved in substrate binding, while Zn1 and Arg-336 are the electrophiles that polarize the scissile carbonyl bond [85]. The cocatalytic site of *E. coli* alkaline phosphatase serves as a precedent for a ligand being involved in catalysis [7, 88]. In this case Ser-102 forms a phosphate intermediate during catalysis.

The structure of *Aeromonas proteolytica* aminopeptidase has been determined in its free state [5] and in a complex with a hydroxamate inhibitor [80]. The two zincs are separated by 3.5 Å. In this case one zinc, Zn1, is coordinated to both carboxylate oxygens of Glu-152, the $N\varepsilon2$ of His-256 and a bridging water and a bridging carboxylate $O\gamma1$ of Asp-117 [5]. The second zinc, Zn2, is bound to both carboxylate oxygens of Asp-179, the $N\varepsilon2$ of His-97, and the bridging water molecule and carboxylate $O\gamma2$ of Asp-117. Thus the cocatalytic zinc sites of AAP and BLAP differ in several details. AAP uses both an Asp carboxylate and a water molecule as bridging ligands while BLAP uses the carboxylate oxygens of a Glu residue and one oxygen of an Asp residue. No bound water is found in BLAP. His ligands are bound to both sites in AAP, whereas no His residues are involved in BLAP. On the other hand, BLAP uses a Lys residue to bind Zn at the tightly bound site. These combinations of ligands as well as the difference in interatomic distance of the two zincs could lead to differences in the charge on the zinc that in turn could influence catalysis.

The results of a p-iodo-D-phenylalanine hydroxamate inhibitor complex with AAP have led to the suggestion of a mechanism for the AAP-catalyzed hydrolysis of peptides [80]. Both the hydroxyl and carbonyl oxygen atoms of the hydroxamate bind to Zn1, while only the hydroxyl oxygen, albeit at a shorter distance, binds to Zn2. The distance between the zinc increases to 3.7 Å in the complex. In addition, the carboxylate oxygen of Glu-151 hydrogen bonds the hydroxylamine nitrogen in a manner seen in several of the matrix metalloproteinases [17, 18, 20, 21] and thermolysin [89]. The similarity of these structures has led the authors to propose Glu-151 as the general base catalyst in AAP [80].

Future mechanistic studies of zinc exo- and endo-proteases will probably benefit from a combination of mutagenic, structural and transient state kinetic studies. In this manner it may be possible to sort out which amino acids are critical to catalysis and to test various mechanistic proposals.

8
Time-resolved XAFS:

Our XAFS, pH and inhibitor studies of ZnCPD indicate it should be possible to examine intermediates in catalysis by XAFS if suitable devices can be made to both mix the enzyme and substrate and monitor the reaction between them at low temperatures. We have therefore begun the construction of a high-economy, low-temperature, stopped-flow, LTSF, with design principles similar to the one used for cryospectrokinetic studies [90, 91] and modified features for XAFS.

Our initial time-resolved XAFS efforts have used carbobenzoxy-sarcosyl-L-phenylalanine, Z-Sar-L-Phe, a substrate that is slowly hydrolyzed by CPD A with a k_{cat}/K_m value of $10^3\,M^{-1}\,sec^{-1}$ at 25 °C about ~ 100 times smaller than for Z-Gly-L-Phe but 100 times larger than for Gly-L-Tyr, the pseudo-substrate used for the X-ray crystallographic studies [53]. At 5 °C the XAFS spectrum of the ES complex of Z-Sar-Phe and ZnCPD ($S_T/E_T = 2/1$) is characterized by near edge features which have two maxima with an intensity ratio of 0.994 and a shoulder at 9683 eV [53] (Fig. 10). Time-resolved XAFS demonstrates the conversion of this intermediate into an XAFS spectrum that is characteristic of the free enzyme plus L-Phe. The near-edge spectrum of this species has two maxima with an intensity ratio of 1.049 and no apparent shoulder at 9683 eV. The plot of the normalized difference in the edge spectra versus time follows first-order kinetics after a lag phase of about one hour. This would be expected for an enzyme-catalyzed reaction in which the substrate is ≥ 20 fold above its K_m and the ratio of S_T/E_T is > 1. At 5 °C the reaction has been slowed sufficiently so that a steady-state phase is now observed since the enzyme is saturated with substrate and the excess substrate is being turned-over. The concentration of the ES complex does not change until S_T becomes less than E. The decay in the concentra-

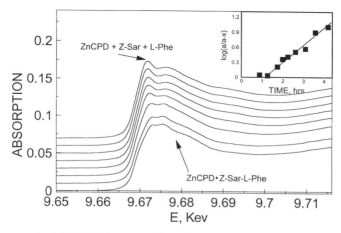

Fig. 10. Time-resolved XAFS of the conversion of Zn CPD · Z-Sar-L-Phe to free enzyme plus products at 5 °C [53]

tion of the ES complex should follow first-order kinetics under these conditions. The results at 5 and 15 °C show that the half-life of the intermediate at 15 °C is increased 2.5-fold to 50 min by decreasing the temperature by 10 °C. The structure of the metal coordination sphere of this intermediate in solution can therefore be readily determined at 5 °C if XAFS spectra are obtained at a higher ratio of substrate to enzyme concentration (e.g. $S_T/E_T = 10/1$). In addition, if lower temperatures can be reached, the intermediates of higher turnover substrates can be examined. Moreover, by mixing the enzyme and substrate at such low temperatures the intermediate in the observation cell can be frozen in liquid nitrogen and the structure of the frozen state intermediate determined at 150 K. In this manner the solution and the frozen states of an ES complex can be readily compared. Such studies should be useful in deciphering the role of the zinc atom in catalysis.

9
Zinc Chemistry Pertinent to Biological Function

Zinc has acquired its prominence because its chemistry is so versatile [1, 3]. It has a remarkably adaptable coordination sphere that allows it to accommodate a broad range of coordination numbers and geometries when forming complexes with N, O and S ligands. Four, five and six coordinate complexes are the ones encountered most frequently in metalloenzymes and other proteins, with regular or distorted geometries ranging from tetrahedral to trigonal bipyramidal, square pyramidal and octahedral. The flexibility in geometry is critical to the role of zinc in catalysis, where expansion or contraction of the metal coordination sphere probably occurs during the formation and breakdown of intermediates. The fact that zinc can exist both as the aquo- and hydroxy-metal complex at pH values near neutrality in addition to its Lewis acid properties is also germane to its role in catalysis. Finally, zinc has a filled d-shell and thus does not undergo oxidation or reduction in contrast to some of its neighboring transition metal ions, such as Cu and Fe, whose oxido-reductive properties are essential to their function. Thus zinc is a stable metal ion species in a biological medium whose redox potential is in constant flux.

10
References

1. Vallee BL, Auld DS (1990) Biochemistry 29:5647
2. Vallee BL, Falchuk KH (1993) Physiol Rev 73:79
3. Vallee BL, Auld DS (1993) Acc Chem Res 26:543
4. Mathews BW (1988) Acc Chem Res 21:333
5. Chevrier B, Schalk C, D'Orchymont H, Rondeau J-M, Moras D, Tarnus C (1994) Structure 2:283
6. Lovejoy B, Cleasby A, Hassell AM, Lingley K, Luthe MA, Weigl D, McGeehan G, McElroy AB, Drewry D, Lambert MH, Jordan SR (1994) Science 263:375
7. Vallee BL, Auld DS (1993) Biochemistry 32:6493
8. Vallee BL, Auld DS (1990) Proc Natl Acad Sci USA 87:220
9. Brändén CI, Jörnvall H, Eklund M, Furugren B (1975) In Enzymes Boyer PD ed Academic Press New York Vol 113rd ed p 103
10. Honzatko RB, Crawford JL, Monaco HL, Ladner JE, Edwards BFP, Evans DR, Warren SG, Wiley DC, Ladner RC, Lipscomb WN (1982) J Mol Biol 160:219
11. Soler D, Nomizu T, Brown WE, Chen M, Ye Q-Z, Van Wart HE, Auld, DS (1994) Biochem Biophys Res Commun 201:917
12. Soler D, Nomizu T, Shibata Y, Brown WE, Auld DS (1995) J Protein Chem 14:511
13. Auld DS (1992) Faraday Discuss 93:117
14. Jiang W, Bond JS (1992) FEBS Lett 312:110
15. Bode W, Gomis-Rüth F-X, Stöcker W (1993) FEBS Lett 331:134
16. Stöcker WS, Grams F, Baumann U, Reinemer, P, Gomis-Rüth, F-X, McKay DB, Bode W (1995) Protein Sci 4:823
17. Borkakoti N, Winkler FK, Williams DH, D'Arcy A, Broadhurst MJ, Brown PA, Johnson WH, Murray EJ (1994) Struct Biol 1:106
18. Spurlino JC, Smallwood AM, Carlton DD, Banks TM, Vavra KJ, Johnson JS, Cook ER, Falvo J, Wahl RC, Pulvino TA, Wendoloski JJ, Smith DL (1994) Proteins Struct Funct Genet 19:98
19. Grams F, Reinemer P, Powers JC, Kleine T, Pieper M, Tschesche H, Huper R, Bode W (1995) Eur J Biochem 228:830
20. Bode W, Reinemer P, Huber R, Kleine T, Schnierer S, Tschesche H (1994) EMBO J 13:1263
21. Stams T, Spurlino JC, Smith DL, Wahl RC, Ho TF, Qoronfleh MW, Banks TM, Rubin B (1994) Nature Struct Biol 1:119
22. Becker JW, Marcy AI, Rokosz LL, Axel MG, Burbaum JJ, Fitzgerald PMD, Cameron PM, Esser C, Hagmann WK, Hermes JD, Springer JP (1995) Protein Sci 4:1966
23. Van Doren SR, Korochkin AV, Hu W, Ye Q-Z, Johnson LL, Hupe DJ, Zuiderweg ERP (1995) Protein Sci 4:2487
24. Gooley PR, Johnson BA, Marcy AI, Cuca GC, Salowe SP, Hagmann WK, Esser CK, Springer JP (1993) Biochemistry 32:13098
25. Browner MF, Smith WW, Castelhano AL (1995) Biochemistry 34:6602
26. Sillén LG, Martell AE (1971) Stability Constants of Metal-Ion Complexes Special Report Publ No 25 The Chemical Society London
27. Argos P, Garavito RM, Eventoff W, Rossmann MG, Brändén CI (1978) J Mol Biol 126:141
28. Christianson DW, Alexander RS (1989) J Amer Chem Soc 111:6412
29. Krebs JF, Ippolito JA, Christianson DW, Fierke CA (1993) J Biol Chem 268:27458

30. Kiefer LL, Paterno, SA, Fierke CA (1995) J Amer Chem Soc 117:6831
31. Lesburg CA, Christianson DW (1995) J Amer Chem Soc 117:6844
32 Silverman DN, Lindskog S (1988) Acc Chem Res 21:30
33. Crowder MW, Wang Z, Franklin EP, Zovinka EP, Benkovic SJ (1996) Biochemistry 35:12126
34. Concha NO, Rasmussen BA, Bush K, Herzberg O (1996) Structure 4:823
35. Hooper NM (1994) FEBS Letters 354:1
36. Rawlings ND, Barrett AJ (1995) Methods in Enzymology 248:183.
37. Haeggström JZ, Wetterholm A, Shapiro R, Vallee BL, Samuelsson B (1990) Biochem Biophys Res Commun 172:965
38. Haeggström JZ, Wetterholm A, Vallee BL, Samuelsson B (1990) Biochem Biophys Res Commun 173:431
39. Medina JF, Wetterholm A, Rådmark O, Shapiro R, Haeggström JZ, Vallee BL, Samuelsson B (1991) Proc Natl Acad Sci USA 88:7620
40. Stöcker W, Wolz RL, Zwilling R, Strydom DJ, Auld DS (1988) Biochemistry 27:5026
41. Nakahama K, Yoshimura K, Maumoto R, Kikuchi M, Lee IS, Hase T, Matsubara H (1986) Nucleic Acids Res 14:5843
42. Goldberg GI, Wilhelm SM, Kronberger A, Bauer EA, Grant GA, Eisen AZ (1986) J Biol Chem 261:6600
43. McKerrow JH (1987) J Biol Chem 262:5943
44. Titani K, Hermodson MA, Ericsson LH, Walsh KA, Neurath H (1972) Nature (london) New Biol 238:35
45. Stöcker W, Ng M, Auld DS (1990) Biochemistry 29:10418
46. Dumermuth E, Sterchi EE, Jiang W, Bond J, Flannery AV, Beynon RJ (1991) J Biol Chem 266:21381
47. Bond J, Beynon RJ (1995) Protein Science 4:1247
48. Bode W, Gomis-Rüth FX, Huber R, Zwilling R, Stöcker W (1992) Nature 358:164
49. Gomis-Rüth FX, Kress LF, Bode W (1993) EMBO Journal 12:4151
50. Baumann U, Wu S, Flaherty KM, McKay DB (1993) EMBO Journal 12:3357
51. Auld DS, Zhang K (1995) FASEB J 9:A1326
52. Zhang K, Auld DS (1993) Biochemistry 32:3844
53. Zhang K, Dong J, Auld DS (1995) Physica B 208 & 209:719
54. Zhang K, Auld DS (1995) Biochemistry 34:16306
55. Larsen KS, Zhang K, Auld DS (1996) J Inorg Biochem 64:149
56. Auld DS, Vallee BL (1970) Biochemistry 9:4352
57. Auld DS, Vallee BL (1971) Biochemistry 10:2892
58. Auld DS, Vallee BL (1987) Carboxypeptidase A In: Neuberger A, Brocklehurst K (eds) Hydrolytic Enzymes Elsevier, Amsterdam New York Oxford, p 201
59. Lipscomb WN (1973) Proc Natl Acad Sci USA 70:3797
60. Riordan JF, Hayashida H (1970) Biochem Biophys Res Commun 41:122
61. French TC, Yu N-T, Auld DS (1974) Biochemistry 13:2877
62. Geoghegan KF, Holmquist B, Spilburg CA, Vallee B L (1983) Biochemistry: 22:1847
63. Bicknell R, Schaffer A, Bertini I, Luchinat C, Vallee BL, Auld DS (1988) Biochemistry 27:1050
64. Bertini I, Luchinat C, Messori L, Monnanni R, Auld DS, Riordan JF (1988) Biochemistry 27:8318
65. Auld DS, Bertini I, Donaire A, Messori L, Moratal JM (1992) Biochemistry 31:3840
66. Larsen KS, Auld DS (1989) Biochemistry 28:9620
67. Larsen KS, Auld DS (1991) Biochemistry 30:2614
68. Christianson DW, Mangani S, Shoham G, Lipscomb WN (1989) J Biol Chem 264:12849
69. Latt SA, Vallee BL (1971) Biochemistry 10:4263
70. Galdes A, Auld DS, Vallee BL (1983) Biochemistry 22:1888
71. Geoghegan K F, Galdes A, Martinelli RA, Holmquist B, Auld DS, Vallee BL (1983) Biochemistry 22:2255
72. Auld DS, Galdes A, Geoghegan KF, Holmquist B, Martinelli RA, Vallee BL (1984) Proc Natl Acad Sci USA 81:5041

73. Galdes A, Auld DS, Vallee BL (1986) Biochemistry 25:646
74. Auld DS, Geoghegan KF, Galdes A, Vallee BL (1986) Biochemistry 25:5156
75. Phillips MA, Fletterick R, Rutter WJ (1990) J Biol Chem 265:20692
76. Christianson DW, Lipscomb WN (1989) Acc Chem Res 22:62
77. Auld DS (1987) Acyl Group Transfer-Metalloproteinases In Page M, Williams A (eds) Enzyme Mechanisms Royal Society of Chemistry p 240
78. Cha J, Pedersen MV, Auld DS (1996) Bichemistry 35:15831
79. Prescott JM, Wagner FW, Holmquist B, Vallee BL (1985) Biochemistry 24:5350
80. Taylor A (1993) Trends in Biochemical Sciences 18:167
81. Chevrier B, D'Orchymont H, Schalk C, Tarnus C, Moras D, (1996) Eur J Biochem 237:393
82. Burley SK, David PR, Lipscomb WN (1991) Proc Natl Acad Sci USA 88:916
83. Burley SK, David PR, Sweet RM, Taylor A, Lipscomb WN (1992) J Mol Biol 234:113
84. Kim H, Lipscomb WN (1993) Proc Natl Acad Sci USA 90:5006
85. Kim H, Lipscomb WN (1993) Biochemistry 32:8465
86. Kim H, Lipscomb WN (1994) Advances Enzymology 68:153
87. Van Wart HE, Lin SH (1981) Biochemistry 20:5682
88. Kim EE, Wyckoff HW (1991) J Mol Biol 218:449
89. Holmes MA, Matthews BW (1981) Biochemistry 20:6912
90. Hanahan D, Auld DS (1980) Anal Biochem 108:86
91. Auld DS (1993) Methods Enzymology 226:553
92. Gomes-Ortiz M, Gomis-Rüth FX, Huber R, Avilés FX (1997) FEBS Letters 400:336

Modeling the Biological Chemistry of Vanadium: Structural and Reactivity Studies Elucidating Biological Function

Carla Slebodnick, Brent J. Hamstra, Vincent L. Pecoraro*

Department of Chemistry, University of Michigan, Ann Arbor, MI 48109, USA
E-mail:vlpec@umich.edu

Prior to the 1980s vanadium was primarily studied because of its role as an industrial catalyst. However, with the discovery of its insulin-mimetic properties and its presence and functional role in certain haloperoxidases and nitrogenases, interest in vanadium chemistry from a biological and pharmacological perspective has exploded over the past 20 years. This paper addresses the biological roles of vanadium including (1) the coordination chemistry for biologically relevant oxidation states, (2) theories and discoveries related to the transport, coordination environment, and role of vanadium in both tunicates and *Amanita muscaria* mushrooms which accumulate vanadium in unusually high concentrations, (3) selective cleavage of proteins by photoactivation of vanadium-protein complexes, (4) the insulin-mimetic abilities of vanadyl, vanadate, and peroxovanadate complexes and their proposed mechanisms of action, and (5) the role of structural and functional model compounds in helping to elucidate the structure and function of vanadium in haloperoxidase and nitrogenase enzymes.

Key Words: Vanadium; biological applications; model chemistry; haloperoxidase; nitrogenase.

List of Abbreviations 52

1   Introduction 54

2   Coordination Chemistry of Vanadium 54

2.1   Vanadium(III) 55
2.2   Vanadium(IV) 56
2.3   Vanadium(V) 58

3   Tunicates 61

3.1   Transport and Storage of Vanadium in Tunicates 62
3.2   Oxidation State and Coordination Environment of
        Vanadium in Blood Cells 63
3.3   Method of Vanadium Reduction Within Blood Cells 65
3.4   Biological Role of Vanadium Accumulation 66

4   Amavadin 66

4.1   Amavadin 66
4.2   Model Compounds 68

* Corresponding Author.

Structure and Bonding, Vol. 89
© Springer Verlag Berlin Heidelberg 1997

5   Insulin Mimicry   69

5.1   Insulin and the Insulin Receptor   70
5.2   Vanadium as an Insulin Mimic   71
5.3   Mechanism for Insulin Mimicry by Vanadium   73
5.4   Conclusions   74

6   Photocleavage   74

6.1   Photocleavage by Monovanadate   75
6.2   Photocleavage by Polyvanadates   77

7   Vanadium Haloperoxidase   78

7.1   Vanadium Haloperoxidase   78
7.2   Model Compounds   80
7.3   Vanadium Peroxidases   94

8   Nitrogenase   95

8.1   Nitrogenase   95
8.2   Model Compounds   96

9   Conclusions 101

10   References 103

List of Abbreviations

18-C-6	18-crown-6 ether
A_0	isotropic hyperfine coupling constant
$A_{\parallel}$	parallel hyperfine coupling constant
acac	acetylacetonato
ADP	adenosine diphosphate
AMP	adenosine monophosphate
ATP	adenosine triphosphate
ATPase	adenosine triphosphate phosphatase
BB rats	BioBread Wistar rats
BMOV	bis(maltolato)oxovanadium(IV)
BrPO	bromoperoxidase
Br-SALIMH	anion of {4-[2-(5-bromosalicylidene)aminato]ethyl}imidazole
Br-SHED	dianion of N-(5-bromosalicylideneaminato)-N'(2-hydroxyethyl) ethylenediamine
CAT or cat	catechol dianion
ClPO	chloroperoxidase
CTP	cytidine triphosphate
CysOMe	cysteine methyl ester
cytPTK	cytosolic protein tyrosine kinase
DBC	3,5-di(*tert*-butyl)catechol dianion

DMF	*N,N*-dimethylformamide
DMSO	dimethylsulfoxide
dppe	bis(diphenylphosphino)ethane
ehpg	ethylenebis[(*o*-hydroxyphenyl)glycine]
ENDOR	electron-nuclear double resonance
Enz	enzyme
EPR	electron paramagnetic resonance
ESEEM	electron spin echo envelope modulation
EXAFS	extended X-ray absorption fine structure
FACS	fluorescence activated cell sorting
GTP	guanosine triphosphate
H_2ada	*N*-(2-amidomethyl)iminodiacetic acid
H_2aeida	*N*-(2-aminoethyl)iminodiacetic acid
Hbpg	*N,N*-bis(2-pyridylmethyl)glycine
H_4edta	Ethylenediaminetetraacetic acid
H_3heida	*N*-(2-hydroxyethyl)iminodiacetic acid
H_3hida	*N*-hydroxyiminodiacetic acid
H_3hidpa	L,L-*N*-hydroxyimino-α,α'-dipropionic acid
his	histidine
Hpic	2-picolinic acid
H_2pmida	*N*-(2-pyridylmethyl)iminodiacetic acid
HPO	haloperoxidase
HSALIMH	[4-2-(salicylideneaminato)ethyl]imidazole
H_2SHED	*N*-(salicylideneaminato)-*N'*-(2-hydroxyethyl)ethylenediamine
H_3nta	nitrilotriacetic acid
IPO	iodoperoxidase
IRK	insulin receptor kinase
LMCT	ligand-to-metal charge transfer
L(SH)$_3$	1,3,5-tris((4,6-dimethyl-3-mercaptophenyl)thio)-2,4,6-tris (*p*-tolylthio)benzene
MCD	magnetic circular dichroism
NADH	nicotinamide adenine dinucleotide, reduced form
NADPH	nicotinamide adenine dinucleotide phosphate, reduced form
Naglavin	Bis(*N*-octylcysteinamide)oxovanadium(IV)
NHE	normal hydrogen electrode
NTP	nucleoside triphosphate
p_i	protein bound inorganic phosphate
PEt$_3$	triethylphosphine
PPTP	protein phosphotyrosine phosphatase
PTK	protein tyrosine kinase
R.D.S.	rate-determining step
salen	*N,N'*-bis(salicylidene)ethylenediamine
SCE	saturated calomel electrode
SQUID	superconducting quantum interference device
STZ	streptozotocin
THF	tetrahydrofuran
tpa	tris(2-pyridylmethyl)amine

tris   *N*-tris(hydroxymethyl)aminomethane
UTP   uridine triphosphate
UV   ultraviolet
UV/vis   ultraviolet/visible
v_i or V_i protein-bound inorganic vanadate
VBrPO   vanadium bromoperoxidase
VClPO   vanadium chloroperoxidase
VHPO   vanadium haloperoxidase
VPO   vanadium peroxidase
XANES   X-ray absorption near edge spectroscopy
XAS   X-ray absorption spectroscopy

1
Introduction

With the discoveries of the role of vanadium as an insulin mimic in 1979, and the presence of vanadium in certain haloperoxidases (1983) and nitrogenases (1986), interest in the coordination chemistry and reactivity of vanadium has greatly increased. Vanadium is believed to be a trace element in most higher animals and known to be essential to some organisms, including tunicates and mushrooms (*Amanita muscaria*). Vanadium functions as a phosphate mimic and a photoactivated amidase making it essential to scientists in characterizing phosphate-binding proteins in vitro. In addition, vanadium's role as a phosphate mimic may account for many of the in vivo and in vitro effects of insulin mimicry. Vanadium-dependent haloperoxidases and nitrogenases are currently the only known enzymes which require vanadium for catalytic activity.

This paper aims to provide a summary of structural and reactivity studies conducted to elucidate the role of vanadium in tunicates, *A. muscaria* mushrooms, vanadium insulin mimics, enzyme photocleavage, haloperoxidases, and nitrogenases. A brief review of the stability and coordination chemistry of biologically relevant oxidation states of vanadium is also provided for the convenience of the reader. For more detailed reviews of the roles of vanadium in biology, two recent books devoted to the subject are recommended [1, 2].

2
Coordination Chemistry of Vanadium

Vanadium was named after Vanadis, the Norse goddess of beauty, because of its propensity to form compounds with a wide variety of colors [3]. Its range of observed oxidation states is equally impressive, with vanadium complexes in oxidation states ranging from −3 to +5 having been thoroughly characterized by various investigators.

Under biologically relevant conditions, however, the extent of available oxidation states is limited to +3 to +5. For this reason, the following discussion of the coordination chemistry of vanadium will concern itself solely with these three oxidation states. The review by Boas and Pessoa is recommended for

Table 1. Reduction potentials in aqueous solution

Reduction reaction	E° (V vs. NHE)
$O_2 + 4H^+ + 4e^- \rightarrow 2H_2O$	1.229
$VO_2^+ + 2H^+ + e^- \rightarrow VO^{2+} + H_2O$	0.991
$VO^{2+} + 2H^+ + e^- \rightarrow V^{3+} + H_2O$	0.337
$V^{3+} + e^- \rightarrow V^{2+}$	−0.225

further information on the coordination chemistry of the lower oxidation states of vanadium [4].

2.1
Vanadium(III)

As Table 1 shows, vanadium(III) is the lowest thermodynamically stable oxidation state for vanadium in aqueous solution. In the presence of oxygen, the blue $[V(H_2O)_6]^{3+}$ ion is readily oxidized as shown:

$$4V^{3+} + 2H_2O + O_2 \rightarrow 4VO^{2+} + 4H^+ \quad E° = 0.892 \text{ V vs NHE} \quad (1)$$

Strictly anaerobic conditions are, therefore, required in the study of biomolecules containing vanadium(III) and their corresponding models. Vanadium(III) can be generated by reduction in tunicates and this oxidation level is maintained by compartmentalizing the ion into special, strongly acidic cells. In model compounds, vanadium(III) can be generated by the disproportionation of vanadium(IV) under acidic conditions [5].

Complexes of vanadium(III) are usually octahedral, although different geometries are occasionally observed, depending on the types of ligands present in solution. The high acidity of $[V(H_2O)_6]^{3+}$ ($pK_{a1} \approx 2.5$, $pK_{a2} \approx 4.0$) leads to the formation of oxo-bridged dimers and other higher nuclearity clusters in moderately acidic solution, which have been studied extensively by Meier et al. and are believed to maintain the octahedral coordination of vanadium [6]. Oxo-bridged vanadium clusters are also observed in the presence of suitable chelating ligands, which frequently help to stabilize vanadium(III) somewhat with respect to oxidation and prevent the precipitation of vanadium(III) hydroxides and oxides at pH values approaching neutrality ($K_{sp,V(OH)_3} = 4 \times 10^{-35}$) [7, 8].

Methods for the spectroscopic characterization of vanadium(III) under biologically relevant conditions are limited. As a d^2, non-Kramer's ion, vanadium(III) is silent under the normal conditions employed for EPR spectroscopy. Although ligand-field bands are present, UV/visible spectroscopy yields limited information about the distortion of the coordination geometry from octahedral symmetry. The types of ligands bound to vanadium may be inferred from the relative energies of these weak d-d transitions, based on trends established from the spectrochemical and nephelauxetic series. EXAFS spectra can determine the approximate coordination number, relative type of ligand, and precise average metrical parameters of atoms bound to the vanadium(III) ion [9].

2.2
Vanadium(IV)

The oxophilicity of vanadium increases with higher oxidation state. Therefore, the chemistry of vanadium(IV) is dominated by the vanadyl ion, VO^{2+}. Nonetheless, there are examples of biologically relevant "bare" vanadium(IV) ions. Non-oxo vanadium(IV) has been found to exist in complexes upon protonation of the vanadyl ion by HCl under anhydrous conditions [10], in the presence of hydroxylamine ligands [11], and in the presence of catecholate ligands [12]. Despite these cases, it is the vanadyl ion which remains the focal point of the biochemistry of vanadium(IV) (with the notable exception of amavadin). Consequently, the vanadyl ion and its properties will be the focus of this discussion.

Vanadyl ions are unstable with respect to aerobic oxidation in aqueous solution, as shown in Eq. 2:

$$4\,VO^{2+} + 2\,H_2O + O_2 \;\rightarrow\; 4\,VO_2^+ + 4\,H^+ \quad E° = 0.238\,V \text{ vs NHE} \qquad (2)$$

However, the rate of oxidation is slow enough to be negligible in slightly acidic solution, especially in the presence of suitable ligands. Therefore, anaerobic conditions are not always necessary for the study of vanadium(IV) complexes. Furthermore, reducing agents such as glutathione and NADH often provide sufficient reducing equivalents to stabilize appreciable quantities of VO^{2+} in vivo.

Complexes of VO^{2+} generally adopt one of two coordination geometries due to the strong trans influence of the oxo moiety of the vanadyl ion. Either a 5-coordinate square pyramidal geometry, in which the vanadyl oxygen occupies the apex of the pyramid [13], or a 6-coordinate distorted octahedron with the sixth ligand weakly coordinated trans to the vanadyl oxygen are routinely observed (Fig. 1) [14]. Under some circumstances, trigonal bipyramidal geometries have been noted for 5-coordinate vanadium complexes [15]. Chelation greatly increases the solubility of the vanadyl ion at neutral pH values, where the precipitation of insoluble $VO(OH)_2$ becomes an obstacle to its incorporation into biochemical environments ($K_{sp,VO(OH)_2} = 3 \times 10^{-24}$).

The d^1 configuration and electronic structure of the vanadyl ion provide the potential for gleaning a great deal of information from spectroscopic studies of its complexes. Figure 2 shows the molecular orbital diagram for what are essentially the metal-centered orbitals of VO^{2+} as determined by Ballhausen and Gray [16]. The vanadyl ion's lone d electron resides in the nonbonding $b_2(d_{xy})$ orbital, which is directed away from all of the ligands, and the π^* orbitals are

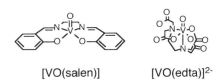

[VO(salen)]   [VO(edta)]²⁻

Fig. 1. Representative VO^{2+} complexes

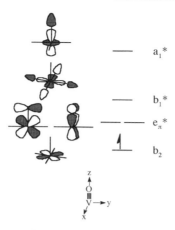

Fig. 2. Molecular orbital diagram for the vanadyl ion

unfilled, so that the vanadium-oxygen bond in VO^{2+} may be formally conside-red to be a triple bond, though in practice it is rarely depicted as such.

Three ligand-field absorption bands are expected in the electronic spectra of VO^{2+} complexes, but only the lower energy $b_2 \rightarrow e_\pi^*$ and $b_2 \rightarrow b_1^*$ bands are usually observed as the $b_2 \rightarrow a_1^*$ transition is usually obscured by the intense LMCT bands in the ultraviolet region of the spectrum. The value of $10Dq$ can be obtained directly from the energy of the $b_2 \rightarrow b_1^*$ transition, which yields an estimate of the average ligand-field strength of the ligands coordinated to VO^{2+} [16].

The S = 1/2 vanadyl ion is well suited for EPR studies. Since over 99% of va-nadium is present as ^{51}V, isotopic enrichment and/or deconvolution of spectra from multiple isotopes is unnecessary. With an unusual nuclear spin (I = 7/2), vanadium(IV) EPR spectra are distinct from those of other transition metals which may be present in biological systems, and are diagnostic for the presence of vanadium(IV). Furthermore, the electronic relaxation rate for VO^{2+} is relativ-ely slow when compared to other transition metal ions. These factors combine to make vanadium(IV) EPR spectra simple to obtain. Eight-line isotropic EPR spectra are readily observed in room temperature solutions of the vanadyl ion and its complexes, and at 77 K, narrow linewidths are observed, enabling the parallel and perpendicular contributions to the spectra to be easily identified and distinguished.

The additivity relationships first put forth by Wuthrich, and expanded by Chasteen and others, constitute an important tool for the interpretation of EPR spectra of the vanadyl ion [17–19]. The observed values for A_0, $A_{\parallel}$, and $g_{\parallel}$ reflect the ligand environment in the equatorial plane of VO^{2+}, and by using the values for individual ligands tabulated in several sources [18, 19], reasonable predic-tions for ligands equatorially bound to VO^{2+} may be made by using the equa-tion:

$$A_{0(calc)} = \sum_i n_i A_{0i} \qquad (3)$$

where i is summed over all types of donors present, n_i (1 to 4) is the number of donors of a given type, and A_{0i} (or $A_{\parallel i}$ or $g_{\parallel i}$) is the contribution to A_0 (or $A_\parallel$ or $g_\parallel$) expected from a single donor of a given type. Of course, meaningful predictions can only be made if some knowledge of potential ligands which may be present is used to guide the choice of ligands used in the calculation.

Unlike some other transition metals, superhyperfine couplings between ligand nuclei and vanadium(IV) are not generally observed in the EPR spectra of VO^{2+} complexes. This is a direct consequence of the fact that the unpaired spin on the vanadium atom is localized in the nonbonding d_{xy} orbital, which leads to couplings between the unpaired spin on vanadium and ligand nuclei which are quite small. Both ENDOR and ESEEM experiments have been effectively used to observe these couplings in a number of biological and nonbiological vanadium complexes. The review by Eaton and Eaton discusses the fundamentals of these two techniques and provides several examples of their application [20].

2.3
Vanadium(V)

Vanadium(V) is the most thermodynamically stable oxidation state of vanadium in aqueous solution under aerobic conditions. As is the case for vanadium(IV), the oxophilicity of high-valent vanadium is the dominant force in determining its chemical behavior.

The chemistry of vanadium(V) is largely dependent on the properties of the various vanadate ions which are found in aqueous solution. Figure 3 shows the different types of vanadate ions which are stable in aqueous solution. At high pH values, or at lower pH under dilute conditions, the monomeric vanadate ion (VO_4^{3-}) in its various protonation states is predominant. At higher concentra-

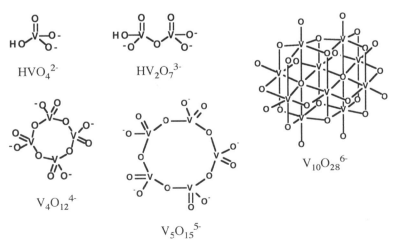

Fig. 3. Structures of the various vanadate ions found in aqueous solution

tion, as the pH is lowered, the dimeric, tetrameric, and pentameric vanadates become increasingly prevalent, such that at pH 7, unligated vanadium (V) in solution exists as a mixture of these four forms, along with a small amount of the decavanadate ion ($V_{10}O_{28}^{6-}$), which, in its various protonation states, is the most abundant form of vanadium(V) found in weakly acidic aqueous solution. In strongly acidic solution, the monomeric dioxovanadium(V) (VO_2^+) ion is the only observed vanadium(V) species. Numerous studies of the relative abundance and interconversion of the different vanadate species at different pH values have been carried out, and the results of these studies, including distribution diagrams, are given in a number of sources [21–23].

It should be noted that the structural similarities between vanadate and phosphate ions lead to similar chemical effects on biological systems. It is commonly believed that ascidians, for instance, accumulate vanadium(V) using the same anion-transport mechanisms by which phosphate is assimilated [24]. The effects of the substitution of vanadate for phosphate in a number of different contexts have been studied; the reader is referred to the reviews by Gresser and Tracey and Stankiewicz et al. for more details [25, 26].

The dioxovanadium ion is the structural unit most commonly found in complexes of vanadium(V) with biologically relevant ligands. In general, these complexes tend to be five- or six-coordinate with geometries which are significantly distorted from idealized square pyramidal or octahedral geometries due to the bent structure of the VO_2^+ ion (O-V-O angle: 101–109°) [27–29]. As in the case of the vanadyl ion, the oxo ligands manifest a strong trans influence, and ligands bound opposite to the oxo groups characteristically show markedly longer bond distances and are substantially more labile than ligands bound in the cis positions.

Floriani et al. have demonstrated the ability of the VO_2^+ moiety to react in a manner reminiscent of a carboxylate group [30]. This "carboxylate analogy" explains the formation and interconversion of vanadium complexes containing numerous VO_2^+-derived units, such as the acidic $VO(OH)^{2+}$, "vanadate esters" of the form $VO(OR)^{2+}$, and anhydride-type structures of the type $(LVO)_2O$, all of which may have important biological roles in various environments. Crans and co-workers have demonstrated the ability of vanadate esters of sugars to act as potent kinase inhibitors and as alternative substrates for other organophosphate-dependent enzymes [31, 32]. Another vanadate ester, uridine monovanadate, has been shown to inhibit a number of nucleases [33, 34].

An increasingly common group of complexes that are structurally similar to vanadyl compounds contain the monooxovanadium(V) (VO^{3+}) ion. Strong donors such as phenolates, alkoxides, and thiolates are required to stabilize these compounds against hydrolysis to VO_2^+ complexes [35–38]. Strong phenolate- or thiolate-to-vanadium charge-transfer bands cause these complexes to be deeply colored.

In certain cases, monooxovanadium(V) complexes are sufficiently oxidizing to chemically modify bound ligands. For example, the complex VO(ehpg) undergoes an unusual stepwise oxidative decarboxylation reaction, in which the final product is the vanadium(IV) complex VO(salen) [39]. This reaction is believed to proceed by a mechanism in which the vanadium(V) starting material is reduced by two electrons to vanadium(III) with the loss of a coordinated carboxylate as CO_2, followed by re-oxidation to vanadium(V) by atmospheric dioxygen and a second decarboxylation, resulting in a vanadium(III)-salen complex which is oxidized to yield VO(salen) [40].

An additional feature of the chemistry of vanadium(V), which is similar to that of vanadium(IV), is the formation of complexes of the "bare" V^{5+} ion under certain conditions [12, 41]. With the possible exception of oxidized amavadin, there are no known biological functions for these oxo-deficient compounds.

One of the more important features of vanadium(V) is its ability to bind and activate peroxide. Peroxovanadates may be generated by the addition of hydrogen peroxide to vanadate solutions, resulting in the exchange of oxo ligands for side-on bound peroxo ligands. As with vanadate chemistry, peroxovanadate chemistry varies with the pH of the solution. In basic solutions, the unstable tetraperoxovanadate $V(O_2)_4^{3-}$ ion may be formed. As the pH is lowered, triperoxovanadate $(VO(O_2)_3^{3-})$ is found in weakly basic solution, diperoxovanadate $VO(O_2)_2^{-}$ predominates in neutral solutions, and the monoperoxovanadate ion $VO(O_2)^{+}$ is the major peroxovanadate species in acidic solution [42].

Complexes of $VO(O_2)_2^{-}$ and $VO(O_2)^{+}$ exhibit distorted pentagonal bipyramidal geometries as illustrated in Fig. 4 [43, 44]. In these complexes, the peroxo ligands have similar structural effects on the structure as do oxo ligands (shortened vanadium-oxygen bond lengths, *trans* influence on other ligands, etc.) so that dioxovanadium complexes and monoperoxovanadium complexes, for instance, are often quite similar structurally and may be considered to be related to each other by the replacement of an oxo ligand by a peroxo ligand, and vice versa. Spectroscopically, however, peroxovanadium complexes are quite distinct, and are easily identified by the presence of peroxo-to-vanadium charge transfer bands (for $VO(O_2)_2^{-}$ complexes, $\lambda_{max} \approx 350$ nm, $\varepsilon \approx 600$ $M^{-1}cm^{-1}$; for $VO(O_2)^{+}$ complexes, $\lambda_{max} \approx 450$ nm, $\varepsilon \approx 300$ $M^{-1}cm^{-1}$) [42].

Peroxovanadates and their complexes have been utilized as oxidants for a wide variety of substrates. The recent review by Butler thoroughly discusses developments in this area; it and the references contained therein are recommended for more details on the reactivity and mechanisms of peroxovanadium-mediated oxidations [42].

Since vanadium(V) complexes are diamagnetic, NMR spectroscopy is a very effective tool for the determination of the structures of vanadium(V) complexes in solution. ^{1}H and ^{13}C NMR provide information about the coordination

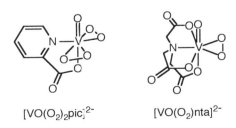

$$[VO(O_2)_2pic]^{2-}   [VO(O_2)nta]^{2-}$$

Fig. 4. Representative peroxovanadium complexes

of ligands to vanadium, not only by giving an indication of the relative symmetry of a complex through the number of peaks observed, but also by observing the presence and magnitude of coordination-induced shifts in peaks in the spectra of complexes as opposed to the spectrum of the free ligand. These techniques have been used by several investigators to examine complexes of vanadium(V) with several ligands similar to those which may be found in biological systems [35, 36, 45, 46].

In addition, ^{51}V NMR can also be easily used to characterize the speciation and coordination environment of vanadium complexes in solution. ^{51}V is present in high abundance (vide supra), and is actually one of the more receptive nuclei for NMR (relative receptivity = 0.383; ^{1}H = 1; constant field). In fact, when the relative abundance of elements is taken into account, ^{51}V NMR spectra are at least 2 orders of magnitude easier to obtain than ^{13}C spectra, even when the small quadrupole moment of ^{51}V is taken into account.

^{51}V NMR spectra are quite sensitive to changes in the coordination environment of vanadium. Along with the linewidth variations, which give some idea of the symmetry of the molecule or metal binding site in question, shifts of hundreds of ppm can be observed depending on the types of ligands bound to vanadium(V). This was perhaps most dramatically illustrated in a series of monooxovanadium(V) phenolate complexes studied by Cornman et al., which spanned a chemical shift range of 604 to –232 ppm (vs neat VOCl$_3$) [47].

Typically, the chemical shifts for biologically relevant vanadium(V) compounds lie well upfield of VOCl$_3$. Vanadates and dioxovanadium complexes are usually found in a chemical shift range between –480 and –600 ppm, with values further upfield generally corresponding to complexes whose coordination environments are more oxygen-rich. In keeping with this trend, peroxovanadium complexes are generally found farther upfield than their dioxovanadium counterparts, with monoperoxovanadium complexes having chemical shift values between –550 and –770 ppm, and diperoxovanadium complexes yielding peaks between –700 and –800 ppm. Thus, different types of vanadium complexes in solution can be identified and distinguished by ^{51}V NMR, and, in conjunction with other techniques, a very accurate picture of the speciation and chemical activities of vanadium(V) in a given solution can be obtained. An important illustration of the use of ^{51}V NMR in studying the interactions of vanadate with organic ligands is the study by Crans and Shin which showed the ability of vanadate to bind to several buffers commonly used in biochemical studies of vanadate-protein interactions.[48] Reviews by Howarth, Rehder, and Crans provide a more detailed background and numerous examples of the use of ^{51}V NMR to examine the speciation and structure of vanadium(V) ions and their complexes [22, 49, 50].

3
Tunicates

Ascidians, also referred to as sea squirts or tunicates because of their characteristic ability to squirt water when touched and their hard colorful tunics, are invertebrate filter-feeding marine organisms. In 1911, Henze discovered that ascidians accumulate high levels of vanadium in their blood cells [51]. He later noted that the vanadium is primarily V(III) and that lysed cells are very acidic [52, 53]. This marked the beginning of extensive scrutiny to understand the role of vanadium in ascidians. Recently, a marine fan worm has been discovered, *Pseudopotamilla accelata*, that also accumulates vanadium in a manner similar to the sea squirts [54]. Ascidians are classified into three suborders depending on the type and amount of metal they accumulate [55]. Aplousobranchs and phlebobranchs accumulate V(IV) and V(III), respectively, while stolidobrachs accumulate Fe(II). The function of vanadium in ascidians remains elusive. Research efforts have concentrated mainly on understanding the storage and transport of vanadium, determining the vanadium oxidation state (+3 or +4), and determining the mechanism of reducing V(V) from sea water [24, 56–59].

3.1
Transport and Storage of Vanadium in Tunicates

Ascidians accumulate vanadium in near molar concentrations [56]. Data obtained from monitoring the uptake of ^{48}V into tunicates shows that vanadium is transported to the blood plasma primarily through the bronchial sacs with some additional absorption into the gastrointestinal tract, and negligible absorption through the tunic [60]. Phosphate inhibition shows vanadate is transported via an anionic transport mechanism [56]. ^{51}V NMR confirms that the majority of the vanadium remains in the +5 oxidation state in the blood plasma [61]. A combination of in vivo and in vitro studies on tunicate blood cells shows that vanadium transport through the cell wall is also via a phosphate transport mechanism [56]. However, an additional mechanism involving storage of vanadium in other tissues and incorporation directly into cells during development may also occur [24, 61]. The vanadium is trapped in the blood cells by reducing it from V(V) (HVO_4^{2-} and $H_2VO_4^-$) to V(III) and/or V(IV). The role of the vanadium once in the blood cells is currently unknown (Sect. 3.4).

At least five different types of blood cells have been identified in ascidians: lymphocytes, stem cells, leucocytes, pigment cells, and vacuolated cells. The vacuolated cells, which are the main storage cells for vanadium, can be further divided into morula cells (bright yellow mulberry-shaped cells with several spherical vacuoles that exhibit pumpkin-colored fluorescence), signet ring cells (green-gray cells with one large vacuole that pushes the nucleus off to the side), and compartment cells (green cells with several vacuoles of different sizes) [62]. Morula cells get their yellow color from hydroxylated polyphenolic tripeptides called tunichromes (Fig. 5) that make up as much as 50% of the dry weight of the cells [63]. The strong fluorescence of tunichrome has been very useful for identifying the morula cells.

Before the characterization of tunichrome, it was assumed that vanadium accumulates in the morula cells, giving them their intense yellow color [59]. However, the results summarized in Table 2 strongly support signet ring cells and compartment cells as the primary sites for vanadium storage in a variety of species of ascidians. In addition, X-ray microanalysis shows that vanadium is found inside the vacuoles of the signet ring and compartment cells [64]. This

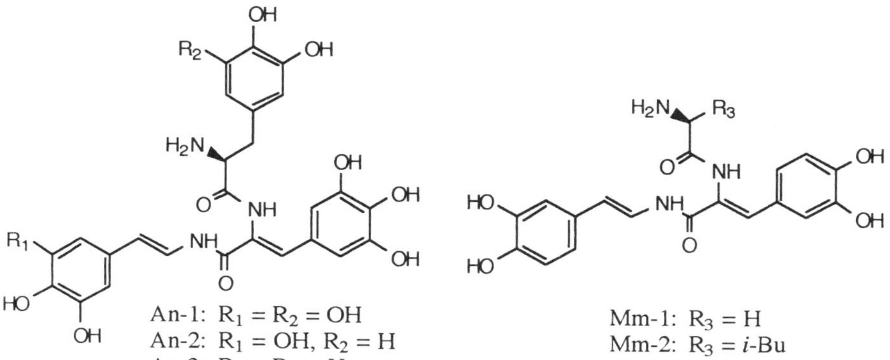

An-1: $R_1 = R_2 = OH$
An-2: $R_1 = OH$, $R_2 = H$
An-3: $R_1 = R_2 = H$

Mm-1: $R_3 = H$
Mm-2: $R_3 = i\text{-Bu}$

Fig. 5. Tunichrome structures

Table 2. Vanadium content in Tunicate Blood Cells

Characterization Technique	Species Studied	Vanadium Content M*	SR*	C*	Ref
X-ray microanalysis	*C. intestinalis, P. mammillata*	trace	yes	yes	[64]
X-ray microanalysis	*A. mentula, A. aspersa*	yes	yes	NA	[74]
cell frac. by density/ neutron analysis	*A. ahodori, A. sydneiensis sameo,*	none	yes	yes	[192, 193]
cell frac. by density/EPR	*A. ahodori*	none	yes	yes	[192]
FACS/AA	*A. ceratodes, A. nigra*	≈ 30%	yes	NA	[65]

* M = morula cell; SR = signet ring cell; C = compartment cell.

breakdown of vanadium in the blood cells is very surprising, because it has been assumed that the redox-active tunichromes play a vital role in vanadium reduction (Sect. 3.1.3), yet it now appears that little of the vanadium is located in the tunichrome-containing morula cells. There is now speculation about the possibility that morula cells develop into signet ring cells and/or compartment cells, suggesting tunichrome may still play a role in vanadium reduction, even though the vanadium concentration is low in morula cells [56, 65].

3.2 Oxidation State and Coordination Environment of Vanadium in Blood Cells

Although the oxidation state of vanadium in the blood cells of ascidians is still somewhat contentious, general agreement appears to be settling on vanadium-(IV) in the aplousobranch suborder and V(III) in the phlebobranch suborder [56, 65]. Most research has concentrated on phlebobranchs, probably partly because V(III) is only found in acidic anaerobic aqueous solutions which are very atypical biological conditions, while V(IV) is a reasonable oxidation state under typical biological conditions (Sect. 2). Therefore a wide variety of experimental data has been required to convince scientists that V(III) is predominant in phlebobranchs. Henze originally determined that blood cells accumulated V(III) by chemical means [52, 53]. Since then, the blood cells of a variety of phlebobranchs have been characterized, often repeatedly, by ^{1}H NMR, EPR, magnetic susceptibility measurements, and EXAFS and the majority of the results still support Henze's original observations.

^{1}H NMR of live blood cells from *A. ceratodes* shows a broad paramagnetically shifted peak at 21.5 ppm corresponding to a labile V(III) aquo complex [66, 67]. The concentration of vanadium was estimated to be approximately 1.3 M in the cell or vacuole where the vanadium is accumulated. However, it was later reported by other researchers that the 21.5 ppm peak may actually be the result of imperfect phasing of the intense free water signal that dominates the spectrum [68]. The EPR spectra of blood cells from *A. nigra* [69], and *A. ceratodes* [67, 70, 71] reveal low concentrations of V(IV). In the study on *A. ceratodes*, the concentration of V(IV) was estimated to be 3–4%, with the remainder V(III). The EPR spectrum fits that of an aquovanadyl complex and it is presumed that the V(III) is also in the form of an aquo complex. However, the reproducibility and validity of this EPR data has also been strongly disputed [68]. SQUID data for *A. nigria* are consistent with a paramagnetic V(III) ion (S = 1) with an anisotropic ground state [72].

V K-edge XAS data from *A. ceratodes* blood cells fit a monomeric V(III) system with no protein or smaller chelated organic molecules [73]. For example, with XANES, the absence of a strong pre-edge absorption common for coordinated oxo complexes (VO^{2+} and VO_2^+), the presence of a weaker pre-edge feature characteristic of ligated sulfate, and the edge energy, are all consistent with a V(III) sulfate complex. The absence of second shell scatters in the EXAFS spectrum shows that organic chelates, such as protein or tunichrome, are not bound and that a monomeric species is present. In addition, the XANES and EXAFS spectra of tunicate blood cells differ only slightly from V(III) in 9 M sulfuric acid [58]. Based on these results, it was concluded that $[V(SO_4)(H_2O)_5]^+$ is the predominant species. Taking into account the possibility of small quantities of VO^{2+}, VO_2^+, and a V(III) or V(IV) triscatechol-type complex when fitting the experimental XANES data, the best fit corresponded to a system that is ~90% $[V(SO_4)(H_2O)_5]^+$, ~10% hexacoordinate tunichrome (modeled as $V(cat)_3^{3-}$ or $V(cat)_3^{2-}$), <5% VO^{2+}, and <2.5% VO_2^+. Considering the data obtained using the different techniques, it now seems clear that the majority of the vanadium is in the form of V(III) aquo complexes. The smaller quantities of vanadium in the morula cells may be coordinated to the tunichrome in the +3 or +4 oxidation state.

If V(III) is the predominant oxidation state, this indicates a strongly acidic anaerobic environment in vanadium-accumulating cells. Henze observed that solutions of lysed cells give a pH of 2.5. Two explanations have been proposed for this unusually low pH: (1) the cells (or subcellular vacuoles) are very acidic or (2) the cells are near neutral pH, but many protons are generated upon cell lysis due to bulk oxidation, ligand exchange, or transition from a very non-polar environment to a polar environment [71]. Cell pHs have been estimated using many different techniques, and there are strong arguments for and against the existence of acidic cells. The size of the paramagnetic shift in ^{1}H NMR [66] and the line width of the parallel $M_I = -7/2$ hyperfine resonance in the EPR [67, 70] are both dependent on acidity and pHs of 1.8–1.9 in the cells where the vanadium is stored have been estimated. Treatment of cells with $BaCl_2$ or $SrCl_2$ also indicates a high acid concentration, thus confirming the presence of high concentrations of sulfate, especially in signet ring cells [74]. Sulfur K-edge EXAFS shows that the sulfur present in high concentrations is in the form of sulfate, as well as sulfonate and disulfide. Broadening of the K-edge peak suggests that sulfate and sulfonate are in a highly acidic form and chelated to vanadium, also pointing to high acid concentrations in the vanadium containing cells [75].

^{31}P NMR chemical shift values [76] and a technique involving equilibration of radioactively labeled markers between the cell vacuoles, the cytoplasm and extracellular fluids with separated cells (morula cells only) [77], gave almost neutral pH values. More recent studies involving lysis of separated cells suggest that the morula cells are neutral or slightly acidic, while the vanadium-containing signet ring cells are very acidic. Calculation of cellular pH based on EPR line widths gives almost neutral pH values for morula cells and a pH of 1.5 for signet ring cells [71, 78]. Assuming that only the vacuoles of the signet ring cells and the compartment cells are acidic, most of the data supporting a neutral cellular pH or an acidic cellular pH are accurate for the conditions of the experiment.

3.3
Method of Vanadium Reduction Within Blood Cells

The method of reduction of V(V) from sea water to V(III) within the cells is unknown. Because the pyrogallol and catechol functional groups found in tunichromes are known redox-active ligands and because the total concentration of tunichrome in the blood cells is of the same order of magnitude as the total vanadium concentration, it is natural to assume that the tunichrome *must* play a role in vanadium reduction. However, the fact that the tunichrome is located primarily in the morula cells and the vanadium primarily in the signet ring and compartment cells, makes this assumption questionable. In addition, even if the tunichrome and vanadium share the same cell during the reduction process, whether the oxidation potential of tunichrome is great enough to reduce V(IV) to V(III) is still doubtful [79].

Currently there is no in vivo evidence which indicates that tunichrome plays any role in vanadium reduction. In vitro studies using tunichromes and catechol or pyrogallol model compounds have shown that they can reduce V(V) to V(IV). With the exception of one report of reduction from V(V) to V(III) in aqueous solution [80] that was later reported to be non-reproducible [81], reduction to vanadium(III) has only been successful in non-aqueous solution. This was accomplished in THF by addition of 2 equivalents of pyrogallol to $VO(acac)_2$, forming the dimeric complex, $V_2(acac)_4\{\mu\text{-}OC_6H_3(OH)\}_2$ [82]. Disproportionation of a V(IV) complex to V(III) and V(V) under acidic conditions in acetonitrile has also been observed suggesting that reduction of V(V) to V(IV), followed by disproportionation, could be a potential route to V(III) formation [5].

Studies in aqueous solution have been performed under acidic, neutral, and basic conditions and consistently yield V(IV) complexes [79, 81, 83]. In one study performed under acidic conditions (pH 2), tunichrome from an iron-accumulating ascidian (*M. manhattensis*) was shown to reduce V(V) to V(IV). All colorimetric assays for V(III) were negative [81]. Studies with tunichromes An-1,2 (Fig. 5) from vanadium-accumulating ascidians were conducted at pH 2 and pH 7 and no detectable quantities of V(III) were formed based on colorimetric analysis [83]. However, for the pH 7 solution, oxidized tunichrome absorbs in the region of interest for the colorimetric analysis, so V(III) formation could not be completely ruled out. XAS on a precipitate formed by the reaction of V(V) with An-1 tunichrome (Fig. 5) in slightly more basic media suggests a small quantity of V(III) was formed (9 ± 5%). In the presence of other potential reducing agents likely to be found in the cell environment, such as glutathione, NADPH, and H_2S, no detectable levels of V(III) are found at pH 7 by XAS or colorimetric analysis. There has been little speculation about other potential reducing agents for V(IV). Thus, the mechanism of formation of V(III) could be due to exogenous reducing agents or disproportionation chemistry, as mentioned earlier.

3.4
Biological Role of Vanadium Accumulation

Over 80 years after Henze first discovered that ascidians accumulate vanadium, many scientists have confirmed most of Henze's original observations using modern, more reliable techniques. The most recent advances have been characterization of the structure of the tunichrome and characterization of contents of the different blood cells. However, so far there is still only speculation regarding the role of vanadium in ascidians. Scientists have suggested the vanadium is there (1) to support anaerobic metabolism by storing highly redox-active metals and ligands, (2) to react with oxygen after injury and produce peroxide as an immunological defense mechanism, and (3) to polymerize tunichrome. Polymerization of tunichrome could serve as a blood clotting mechanism after cell lysis due to injury, as the method for tunic generation, or as an adhesive to fix the ascidians to underwater surfaces [56, 58, 84].

4
Amavadin

4.1
Amavadin

In 1931, Ter Meulen reported that the mushroom *Amanita muscaria* contained unusually high concentrations of vanadium [85]. It was not until 1972, however, that the vanadium-containing compound in this mushroom was isolated by Kneifel and Bayer, and given the name amavadin (or amavadine) [86]. Since that time, several other *Amanita* species have been shown to have high vanadium contents as well, with vanadium concentrations of nearly 400 ppm being observed in some species [87].

Upon the identification of the organic ligand present in amavadin and examination of its IR and EPR spectra, Kneifel and Bayer proposed that amavadin was a 2:1 complex of L,L-*N*-hydroxyimino-α,α'-dipropionic acid (H₃hidpa) with vanadyl ion (Fig. 6a) [88]. The key data involved in the assignment of this structure (along with the determination of the structure of the ligand) were the axial EPR spectrum of amavadin, which is similar to those typically observed for vanadyl complexes (*vide supra*), and an absorption in the IR spectrum at 980 cm⁻¹, which was assigned as a V = O stretching frequency.

During the 1980s, the total synthesis of amavadin was reported by a number of groups, all of whom reported data which appeared consistent with the proposed structure [89, 90]. As studies of amavadin progressed, however, it became clear that the proposed structure did not satisfactorily explain all of its properties. The stability constant determined for amavadin was ten orders of magnitude greater than that for any other divalent ion studied, a fact for which the proposed structure offered no explanation [91]. Furthermore, large-angle X-ray scattering experiments indicated that there were no short vanadium-oxygen distances, as are always observed in vanadyl complexes [92]. The vanadium(IV) complex of the analogous ligand *N*-hydroxyiminodiacetic acid (H₃hida), which lacks only the methyl groups of H₃hidpa, though yielding UV/visible and EPR spectra virtually identical to those of amavadin, contained no 980 cm⁻¹ band in its IR spectrum [92]. An alternative binding mode for the hydroxylamino moieties of H₃hidpa was suggested by Wieghardt and co-workers, who demonstrated that hydroxylamines were capable of side-on coordination to vanadium [93]. In the light of these and other facts, Bayer and co-workers proposed a structure similar to that shown in Fig. 6b for amavadin, in which the hydroxylamino moieties of the ligands are ionized and bound side-on to vanadium, resulting in an eight-coordinate "bare" vanadium(IV) ion [92]. Confirmation of this unusual structural proposal for amavadin was obtained through a crystallographic study of the model compound [V(hida)₂]²⁻, crystallographic and NMR (¹H and ¹³C) studies of the oxidized form of amavadin, and EXAFS studies of its oxidized and reduced forms [11, 41].

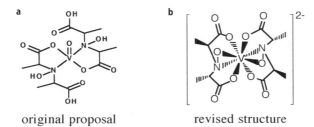

original proposal   revised structure

Fig. 6. a Originally proposed structure of amavadin. **b** Experimentally determined structure of amavadin

The question of what role amavadin may play in vivo remains largely un-answered. Most speculation centers around the ability of amavadin to be re-versibly oxidized by one electron ($E° = 0.53$ V in H_2O, $E° = 0.03$ V in DMSO, both vs SCE)[90]. This has led several groups to propose that amavadin acts as a one-electron redox mediator [90, 94, 95]. Based on the original structural pro-posal, amavadin was believed to act through an inner-sphere mechanism via the highly exchange-labile axial position [95], but with the realization of the correct structure of amavadin, the question of whether amavadin acts as an inner-sphere or an outer-sphere electron-transfer agent (or both) is open to debate. A recent study suggests that amavadin may be used to cross-link pro-teins by oxidizing thiols to disulfides [94], and that this may be part of a defense mechanism similar to the proposed role of vanadium(III) in polymerizing the outer shell of tunicates [84].

4.2
Model Compounds

Much of the work which has led to our current understanding of the structure and potential functions of amavadin has been carried out on the complex $[V(hida)_2]^{2-}$ which offers the advantages that H_3hida is easier to synthesize and isolate in pure form than H_3hidpa and that H_3hida does not present the diffi-culties associated with the presence of stereocenters in H_3hidpa [41, 89, 91, 95].

The crystal structure of $[V(hida)_2]^{2-}$ provided the first evidence that the eight-coordinate structure proposed for amavadin by Bayer was chemically reasonable. In conjunction with the crystallographic studies of oxidized ama-vadin and $[V(hida)_2]^{2-}$ by Garner et al. [41], two points become clear: 1)·that $[V(hida)_2]^{2-}$ and amavadin possess extremely similar structures, and 2) that the oxidation of $[V(hida)_2]^{2-}$ does not induce major structural changes in the com-plex, as shown in Table 3. From this it then follows that amavadin may serve as an efficient electron-transfer agent, since very little reorganization of the coor-dination environment around vanadium is necessary in the electron-transfer process.

Studies of the reactivity of amavadin and $[V(hida)_2]^{2-}$ have been restricted to the use of these complexes in the electrocatalytic oxidation of thiols to disul-fides by Thackerey and Riechel and by Frausto da Silva and co-workers [94, 96]. Table 4 illustrates that the ability of these complexes to mediate thiol oxidation depends on the nature of the substrate, which must contain both thiol and car-boxylic acid functional groups to be oxidized by amavadin. The sole exception to this is the reported electrocatalytic oxidation of the methyl ester of cysteine;

Table 3. Bond lengths for reduced and oxidized $V[(hida)_2]$ complexes

bond type (lengths in Å)	$V[(hida)_2]^{2-}$	$V[(hida)_2]^-$
avg. V-O (carboxylate) bond length	2.067	1.951
avg. V-O (hydroxylamine) bond length	1.975	1.970
avg. V-N (hydroxylamine) bond length	2.003	2.022

Table 4. Substrates tested for electrocatalytic oxidation by amavadin and $[V(hida)_2]^{2-}$

electrocatalytically oxidized	not electrocatalytically oxidized
$(CH_3)_2CH(SH)CH(NH_2)C(O)OH$	$HOCH_2CH(NH_2)C(O)OH$
$HSCH_2CH(NH_2)C(O)OH$	$CH_3SCH_2CH(NH_2)C(O)OH$
$HSCH_2CH(NH_2)C(O)OCH_3$	$HSCH_2CH_2NH_2$
$HSCH_2C(O)OH$	$HSCH_2CH_2CH_3$
$HSCH_2CH_2C(O)OH$	$HSCH_2CH_2CH_2SH$
glutathione (γ-GluC(O)OH-Cys(SH)-	$HSCH_2CH_2SO_3Na$
GlyC(O)OH)	$HSCH_2CH_2OH$

however, the fact that these experiments were carried out in acidic aqueous solution raises the possibility that this is also a reaction of free cysteine, since the
characterization of the oxidation products of this reaction was not reported
[96].

Frausto da Silva et al. have reported that the electrocatalytic oxidation of
thiols by amavadin and $[V(hida)_2]^{2-}$ exhibits saturation behavior, consistent
with an unusual electrocatalytic Michaelis-Menten-like mechanism in which
the complex is oxidized from vanadium(IV) to vanadium(V), followed by the
formation of a thiol-vanadium(V) complex adduct, and the regeneration of the
vanadium(IV) complex by one-electron oxidation and release of the oxidized
thiol [94]. It remains unclear, however, what the structure of this thiol-vana-
dium adduct is. Although there is no evidence to support such a proposal, the
requirement that the thiol must contain a carboxylate to be oxidized, strongly
suggests that the thiol must be coordinated to vanadium through the carboxylate and quite possibly through the thiolate sulfur as well. Such a proposal requires the dissociation of one or more donors of the hida (or hidpa) ligand from
vanadium, and will require further studies of the stability and ligand-exchange
behavior of $[V(hida)_2]^-$ and oxidized amavadin to determine which donors are
likely to be displaced by incoming ligands.

Regardless of the mechanistic details of the electrocatalytic oxidation of
thiols by these complexes, these studies still provide the first evidence that amavadin is capable of interacting and reacting with biologically relevant substrates
and represent an important first step in determining what role amavadin plays
in mushrooms.

5
Insulin Mimicry

Approximately 15 million people suffer from diabetes mellitus in the United
States. These people either cannot produce insulin (insulin-dependent or type I
diabetes) or are unable to respond to insulin (non-insulin-dependent or type II
diabetes). Type I diabetes develops in children because of an insufficient
number of insulin-producing cells in the pancreas, and must be treated with
insulin injections. Type II diabetes occurs in adults when tissue cells no longer
respond to insulin. Two of a number of possible causes for this are a defect in
the insulin receptor or attack of antibodies at the insulin receptor. As a result,

insulin injections are not effective. A strict diet along with administration of drugs that are believed to affect insulin production, glycolysis, gluconeogenesis, glucose transport, and/or fatty acid degradation are common treatments. Some of the many symptoms of diabetes include high blood glucose levels, glucose intolerance, abnormal thirst, overeating, weight gain, and cataracts [97–100].

5.1
Insulin and the Insulin Receptor

Insulin is the primary hormone for controlling most anabolic processes while simultaneously inhibiting many catabolic processes. Insulin stimulates synthesis and inhibits breakdown of glycogen in liver and muscle tissue and stimulates protein synthesis in muscle tissue and triglyceride synthesis in adipose tissue. In addition, insulin triggers glucose, amino acid, and fatty acid uptake into the relevant tissues and it promotes the oxidation of glucose to CO_2 and water. Insulin is a hormone composed of two protein subunits linked by disulfide bonds: an α-subunit of 21 amino acid residues and a β-subunit of 30 residues. Insulin binds to the insulin receptor site on the plasma membrane triggering the metabolic processes discussed above through a cascade of enzymatic reactions [97–100].

Insulin receptor kinase (IRK), the insulin receptor protein, is a transmembrane protein comprised of two α- and two β-subunits linked by disulfide bonds in a β-α-α-β conformation. The α-subunits consist of 719 residues and are situated on the outside of the plasma membrane. These subunits are the site of insulin binding. The β-subunits contain a total 619 residues each, a hydrophobic region that spans the plasma membrane and an intracellular region that acts as a protein tyrosine kinase (PTK) (Fig. 7). IRK is responsible for autophosphorylation of 13 of its own tyrosines as well as phosphorylation of tyrosines on other cellular proteins, triggering the currently poorly understood sequence of reactions responsible for the metabolic effects of insulin binding. Subsequent to autophosphorylation of IRK, the protein remains active regard-

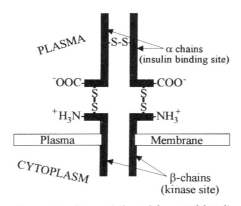

Fig. 7. Schematic of insulin receptor kinase (adapted from Ref. [199])

less of whether or not insulin is bound. Serine and threonine residues of IRK can also be phosphorylated by different kinase enzymes, partially deactivating IRK. Total inactivation is believed to be achieved by a protein phosphotyrosine phosphatase (PPTP) that returns the IRK to its pre-insulin-binding form [97–100].

5.2
Vanadium as an Insulin Mimic

An insulin mimic is a substance that alleviates some or all of the symptoms of diabetes. Alternatives to insulin treatment for diabetes could provide a number of potential advantages. For example, insulin is destroyed during the digestive process and must be administered by injection to be effective. Therefore, a substance that can be administered orally would increase the ease of treatment as well as the availability of treatment to patients in poorer countries. Since type II diabetics cannot be treated by insulin injections, substances that mimic insulin through alternative pathways are necessary. Finally, the study of systems that mimic insulin's activity can help scientists further elucidate the mechanisms of insulin's effects on the body [99].

In 1979, it was discovered that vanadate ($H_xVO_4^{x-3}$) stimulates glucose uptake and oxidation in intact rat adipocytes and as well as glycogen synthesis in diaphragm and liver cells, thus mimicking some effects of insulin [101]. Vanadate and vanadyl have since been shown to stimulate or enhance most effects of insulin, with the exception of increased protein synthesis. Among the physical symptoms alleviated by vanadium are high blood glucose levels, overeating, hyperthyroidemia, excessive thirst, hyperlipidemia, and cataracts. A growing interest in vanadium as an insulin mimic is evident by the number of recent review articles and vol. 153 of *Cellular and Molecular Biochemistry* which are devoted entirely to this subject [97–100].

Initial studies of vanadium as an insulin mimic concentrated on V(V) because it is the most stable and soluble form of vanadium at physiological pH and is easily transported by anion transport mechanisms [24]. The main drawback to using vanadate is that it is the most toxic form of vanadium, causing severe gastrointestinal problems and eventual death if consumed in high enough doses [102]. Vanadate has been studied in vitro and in vivo in the form of $NaVO_3$ dissolved in drinking water. NaCl is often added to the drinking water, helping to reduce the toxic effects. Insulin-mimetic effects of V(V) require 0.2–1 mg/mL concentrations in drinking water. These concentrations are high enough for rats to develop diarrhea and reduce food and fluid intake, resulting in dehydration and possible death [102]. Although V(V) is considered the most stable oxidation state, EPR spectra consistent with vanadyl complexed to reduced glutathione indicate that V(V) is at least partially, and perhaps completely, converted to V(IV) by reducing agents within the cell [103]. This raises questions about whether V(V), V(IV), or both V(V) and V(IV) are mimicking insulin.

Vanadyl has been studied to help address the question of whether V(IV) or V(V) is the active species and because it is less toxic than vanadate. It has been studied both as $VOSO_4$ and in the form of VOL_2, where L represents a mono-

Fig. 8. Coordination compounds used to stabilize the V(IV) oxidation state and improve oral absorption

anionic bidentate ligand (Fig. 8) [103]. Not only does vanadyl mimic many of the effects of insulin, but in a number of studies, normoglycemic levels were observed 13–20 weeks after cessation of vanadyl treatment [104]. This is attributed to the apparent long-term ability of vanadium to enhance responsiveness to insulin. Although $VOSO_4$ is less toxic than vanadate, it still poses problems because minimal required doses are still dangerous and oral absorption is poor [99]. By using the coordination complexes described in Fig. 8, both the water solubility and the lipophilicity can be manipulated to improve oral absorption. Among the most-studied of the vanadyl complexes is BMOV (Fig. 8) [105–107]. This complex requires lower doses (1/3–1/2) of vanadium to be as effective as vanadate and shows less gastrointestinal side effects [107]. BMOV mimics most of the responses of vanadyl and vanadate, except it does not reduce weight gain. Toxicology studies on Naglavin suggest that the doses required for insulin-mimetic effects are still within an order of magnitude of doses resulting in death to all test animals within 4 days [108].

When studying insulin-mimetic effects of vanadate and hydrogen peroxide, it was discovered that these two substances act synergistically forming peroxovanadates that are significantly more effective than vanadate or vanadyl [109, 110]. For in vitro studies of IRK activity, doses approximately 10^{-3} times the concentration are required compared to vanadyl and vanadate. In vivo studies with peroxovanadate complexes also show that significant decreases in blood glucose levels are observed at much lower concentrations than are required for vanadate [111, 112]. Evidence shows that the mechanism of action is different than that of vanadyl and vanadate (vide infra) [99]. In the absence of stabilizing ligands, peroxovanadates decompose in aqueous solution and must be administered by injection for effectiveness. Some of the complexes that have been studied are presented in Fig. 9. Studies of dioxo, monoperoxo, and diperoxo complexes indicate that diperoxovanadates are more reactive than monoperoxovanadates and dioxovanadates are the least reactive. In addition, sterically diperoxo   monoperoxo   dioxo

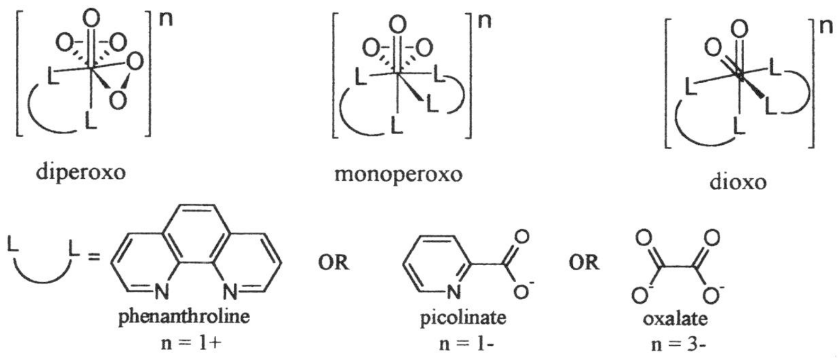

phenanthroline   picolinate   oxalate
n = 1+   n = 1-   n = 3-

Fig. 9. Ligands used to stabilize V(V) peroxo and V(V) dioxo compounds

bulky substituents on the pyridyl or the phenanthroline ligands (Fig. 9) reduce the effectiveness of the peroxovanadates [113].

5.3
Mechanism for Insulin Mimicry by Vanadium

Speculation into the possible mechanism for vanadium's insulin-mimetic effects has concentrated mainly on vanadate's ability to act as a phosphate analog and inhibit PPTPs. Vanadate is a known inhibitor of PPTPs, but not phosphoserine or phosphothreonine phosphatase. This selectivity for tyrosine is very significant because tyrosine represents < 0.05% of phosphorylation sites in the body which could explain why vanadate mimics other biological functions with minimal effect. Although vanadates, vanadyls, and peroxovanadates are believed to inhibit PPTPs, the different mechanisms of response for the vanadium complexes probably result from selectivity for different PPTPs [97, 98, 114].

Vanadate and vanadyl are both believed to inhibit a PPTP selective for a cytosolic PTK (cytPTK) rather than the PPTP that targets IRK [97, 98, 114]. By inhibiting a specific PPTP, vanadate indirectly activates the corresponding cytPTKs. From inhibition studies with IRK, it is clear that vanadate and vanadyl bypass the IRK phosphorylation step or activate a different PTK, since these systems are active regardless of whether IRK is phosphorylated [97, 98, 114]. In recent studies on the inhibition of cytPTK, some, but not all, of vanadium's insulin-mimetic effects are suppressed, suggesting that cytPTK activation may be important, but other, currently unknown mechanisms are also important [114]. Recent studies in cell-free systems where the oxidation state of the vanadium can be controlled, suggest that there are distinct vanadate and vanadyl mechanisms with the different ions showing selectivity for PPTPs that target different cytPTKs [115]. Another theory on vanadate's and/or vanadyl's mechanism of action is that the vanadium ions change the cellular pH or increase Ca^{2+} influx and trigger increased transport of glucose-containing organelles to the plasma membrane and through the cell walls. This process is known to be sensitive to pH changes, among other things [98].

In vitro studies have shown that peroxovanadate complexes inhibit the PPTP that is selective for IRK, causing increased IRK activity. A distinct correlation between % PPTP inhibition and % IRK activity has been observed in hepatic cells, strongly supporting this theory. The mechanism for PPTP inhibition is presumed to be irreversible oxidation through a peroxidase mechanism of a vital cysteine residue located at the site of tyrosine phosphate binding (vide infra) [113].

5.4
Conclusions

There is overwhelming evidence that vanadium in the form of vanadyl, vanadate, and peroxovanadate mimics most of the effects of insulin in both type I and type II diabetic models, although type I BB rats still require insulin injections for survival [116]. With studies to further elucidate the mechanism of action for each of the systems and efforts to develop less toxic coordination complexes, orally administered vanadium may become a useful treatment for type II diabetics and a treatment to be coadministered with insulin to type I diabetics to enhance responsiveness to insulin and reduce the frequency of insulin and/or doses. Very limited studies on humans have shown enhanced sensitivity to insulin in type II and some type I diabetics [117]. However, before vanadium treatment can become reality, much research must concentrate on developing less toxic vanadium complexes.

6
Photocleavage

In 1986, scientists attempted to combine photolabeling and the use of vanadate as a strongly binding phosphate inhibitor to help characterize the motor protein dynein. They discovered that exposure to light at a wavelength of 254 nm resulted in dynein cleavage at one primary site [118]. Since this discovery, dynein, myosin, and adenosine kinase have been the most studied among the many proteins shown to be photocleaved by vanadate [119, 120]. In some cases, multiple vanadium-induced photocleavage sites are present on the proteins and the site that gets cleaved can be controlled by manipulating the concentration of vanadium and ADP in solution.

Dynein and myosin are motor proteins that function as ATPases [121]. Dynein binds to tubulin in cilia and flagella and triggers cell movement and intracellular vesicle transport. The approximately 1500 kDa protein consists of 2 or 3 heavy chains of 428 kDa (α, β, and γ chains), 3 medium chains of 70–120 kDa, and approximately 4 light chains of 15–25 kDa. Each heavy chain has an ATP binding site and a microtubule binding site. Myosin, which regulates muscle contraction, is approximately 540 kDa and consists of 2 heavy chains (230 kDa) and 4 light chains (~20 kDa). A globular head that forms at the NH_2-terminal end of the heavy chains contains the ATP and actin binding sites. This globular head can be easily cleaved to form a 95 kDa protein segment called the S1 fragment. Most myosin photocleavage studies have been done on the smaller and easier to characterize S1 fragment. Adenosine kinase is a much smaller enzyme (~21 kDa) that binds both ATP and AMP, catalyzing phosphate transfer from ATP to AMP [120].

6.1
Photocleavage by Monovanadate

Vanadate is known to inhibit ATPases, presumably by forming stable protein-V_i-MgADP complexes similar to the protein-P_i-ADP transition state [24]. In the presence of vanadate concentrations of ≤ 50 mM and MgADP, a dynein-V_i-MgADP complex is formed that is stable for days or weeks in the absence of light. However, upon exposure to light, photocleavage occurs with $t_{1/2} \approx 7$ min. This cleavage occurs at similar sites in the α and β heavy chains, referred to as the V1 sites, creating fragments of 200 and 228 kDa [122]. Initial experiments were run with light at a wavelength of 254 nm, but non-specific photocleavage was observed along with the specific vanadium-triggered photocleavage [118]. By increasing the wavelength of the light to 365 nm, where vanadate absorbs but the aromatic functional groups of the peptide do not, cleavage at only the V1 site was achieved. At wavelengths above 400 nm, where the vanadate no longer absorbs, no cleavage occurs [122].

It has been concluded that a V_i-MgADP complex binds at the ATP binding site and that this is the site of photocleavage. This is based on the facts that: (1) the V_i-MgADP complex inhibits ATPase activity, (2) substitution of other nucleoside triphosphates (NTPs) for ATP results in parallel rates of hydrolysis and photocleavage (ATP > CTP ≈ UTP > GTP), (3) in the absence of vanadate, or if AMP is substituted for ADP, no photocleavage occurs, (4) the rate of peptide cleavage corresponds to the rate of loss of ATPase activity, (5) exposure to triton X-100 results in similar rate increases for ATPase activity and photocleavage, and (6) denaturation of the peptide results in loss of ATPase activity, V_i-MgADP binding, and photocleavage activity. The photocleaved α and β subfragments can be bound to the flagellar tubules, but mobility is greatly reduced since ATPase activity is necessary for mobility [122, 123].

Introduction of free radical scavengers to a solution containing dynein-V_i-MgADP has no effect on the rate of cleavage. However, the rate of photocleavage is affected by substitution of other divalent cations for Mg^{2+} to form protein-V_i-M^{2+}ADP complexes. Relative photocleavage rates of $Mg^{2+} \approx Zn^{2+} > Ca^{2+} \approx Cu^{2+} > Ni^{2+}$ are observed. No photocleavage is observed with Cr^{2+}, Mn^{2+}, Fe^{2+}, or Co^{2+}, suggesting that these redox-active metals are very near the vanadium, quenching the active chromophore via an inner sphere charge transfer mechanism. From these studies it has been concluded that the mechanism for photocleavage involves a ligand-to-metal charge transfer followed by modification of the peptide at the V_i-MgADP (ATP) binding site, ultimately resulting in peptide photocleavage [122]. The peptide sequence of the β-chain of urchin sperm flagella has been determined and the proposed site of cleavage is at the alanine of a phosphate binding loop (Table 5) [124]. This conclusion is based strongly on data obtained from myosin and adenylate kinase (vide infra).

Table 5. Homologous sequences for ATP binding proteins [194]

Dynein (urchin sperm flagella, β-chain)	G	P	A	G	T	G	K	T	
Myosins	G	E	S	G	A	G	K	T	
Adenylate kinases	G	X	P	G	X	G	K	G	T
ATP synthases (β)	G	G	A	G	V	G	K	T	V
ATP synthases (α)	G	D	R	Q	T	G	K	T	A/S
Thymidine Kinases	G	X	X	G	X	G	K	T	T
Phosphoglycerate Kinases	A/V	X	X	G	G	A/S	K	V	X

Photocleavage studies on the smaller, well-characterized S1 fragment of myosin have proven vital in further probing the mechanism of vanadium-induced photocleavage. Using most of the experiments described above for dynein, it has been shown that V_i-MgADP also binds strongly to the ATP binding site of myosin ($t_{1/2} \approx 3-4$ days) and specific photocleavage occurs at the phosphate binding site upon exposure to light at 300–400 nm, creating subfragments of 21 and 74 kDa. Further studies on the S1 fragment of myosin have shown that S1 photocleavage is a 2-step process involving photomodification of the residue located at the ATP binding site prior to photocleavage. The photomodified intermediate is isolated by binding V_i-MgADP to S1 and removing excess vanadate prior to UV-irradiation. Failure to remove excess vanadate or addition of more vanadate to a solution of the photomodified protein results in photocleavage [125].

The UV spectrum of the photomodified protein shows a pH-dependent absorbance increase at 280 nm compared to the unmodified protein. After eliminating aromatic amino acids as the possible site for photocleavage and conducting pH-dependent studies, it was concluded that the 280 nm absorbance increase at higher pH is an enolate, corresponding to a serine-aldeyhde at neutral and acidic pH. Reduction of the oxidized serine back to an alcohol and specific radiolabeling of this residue was achieved using NaB^3H_4. Cleavage of the radiolabeled peptide with trypsin followed by amino acid analysis shows that the modified serine, located 21 kDa toward the COOH-terminus of a 23 kDa trypsin-digested segment of the S1 fragment (Fig. 10), is the only radiolabeled amino acid. The oxidized serine corresponds to X_2 in a peptide sequence of the form $G-X_1-X_2-X_3-X_4-G-K-G/T$ that is common to most ATP-binding proteins (Table 5) [126, 127].

Studies on chicken muscle adenylate kinase have been very useful in revealing the sites of photocleavage and in characterizing the photocleavage products. Inhibition by ATP, but not AMP was observed, leading to the conclusion that cleavage is at the ATP-binding site of adenylate kinase. Like myosin, the X_2 amino acid of the $G-X_1-X_2-X_3-X_4-G-K-G/T$ phosphate binding loop is the site of photocleavage. However, in adenylate kinase the residue is proline rather than serine, suggesting that the location of the amino acid in the phosphate-binding loop determines the site of photocleavage, rather than the identity of the amino acid. Irradiation in the presence of 0.25–1.5 mM vanadate and ADP results in formation of a 19.5 kDa peptide and a small peptide of ca. 20 amino acids. Based on the lack of correlation between the percentage of enzyme activity and the

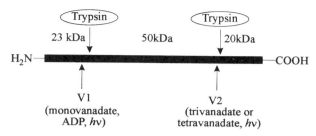

Fig. 10. Sites of vanadium-catalyzed photocleavage of the myosin heavy chain S1 fragment

percentage of cleavage, with a peptide that is 55–60% photocleaved showing <5% activity, it is assumed that the enzyme is photomodified and inactivated prior to cleavage. Presumably, the proline is oxidized and then decarboxylated, forming γ-aminobutyric acid, which has been characterized by sequencing the cleaved segment [128].

6.2
Photocleavage by Polyvanadates

Introduction of 0.05–1 mM vanadate to the dynein heavy chains [129] or the S1 fragment of myosin [130] results in photocleavage at two distinct sites that are independent of ADP concentration. These sites include the V1 site already addressed and a new site called the V2 site. Selective cleavage at the V2 site can be achieved by binding Vi-M^{2+}ATP (M = Co, Mn) to quench photocleavage at V1 (vide supra) [130]. Cleavage at the V2 site is unaffected by free radical scavengers or substitution of redox-active metals for Mg^{2+}, but dynein cleavage is strongly inhibited by NTPs [129]. Subsequent to photocleavage, photoaffinity analogs of NTPs no longer bind [131]. In the presence of tris buffer, photocleavage inhibition is observed and an EPR signal develops at a rate corresponding to the rate of photocleavage. Based on these observations, it is concluded that V(V) is reduced to V(IV) by photomodification and/or photocleavage of both peptide and tris. Tris buffer then binds and stabilizes the reduced form. In the absence of tris, O_2 immediately reoxidizes the vanadium, completing the catalytic cycle [129, 130].

Under the conditions of the experiments, monomeric, dimeric, tetrameric, and pentameric vanadate are the major species present in solution and trivanadate is I, kely to be present in trace amounts. Rate dependence studies of vanadium concentration show a sigmoidal dependence with 2.7 sigmoidicity, suggesting that trimeric vanadate is the species that binds and triggers dynein and myosin photocleavage [129, 132]. Line broadening of the tetrameric peak in ^{51}V NMR studies implies that tetravanadate is interacting most strongly with myosin [130]. Based on these results, it is assumed that trimeric and/or tetrameric vanadate bind at an additional phosphate-binding site in dynein and myosin. In actin-bound myosin, cleavage at the V2 site is inhibited. It is possible that ATP binding at this site may be responsible for regulating the myosin-actin interaction [129, 132]. The V2 site in myosin is about 20 kDa from the COOH-terminus of the S1 chain, corresponding to the positively charged, lysine-rich, actin-binding site of myosin (Fig. 10) [130].

In addition to dynein and myosin, peptide photocleavage by monomeric and/or oligomeric vanadate has also been observed in the proteins ribulose 1,5-bisphosphate carboxylase/oxygenase, isocitrate lyase, aldolase, phosphofructo-kinase, and Ca^{2+}-ATPase. With ribulose 1,5-bisphosphate carboxylase/oxygenase and isocitrate lyase, NaB^3H_4 has been used to reduce stable intermediates and implicate a serine at the site of cleavage [120].

The cleavage of peptides by vanadate has been useful in the characterization of ATP-binding proteins. By expanding studies to include other metals and other proteins, metal-catalyzed photocleavage could become a standard technique in protein characterization. Dynein-catalyzed photocleavage by Fe (III) and Rh (III) has already been observed, with a selectivity for different sites than vanadate [133], further supporting the potential versatility of this technique in peptide characterization.

7
Vanadium Haloperoxidase

7.1
Vanadium Haloperoxidase

Recent advances in the biology and enzymology of vanadium haloperoxidases (VHPOs) are addressed fully in the chapter in this book by Alison Butler and co-workers. Therefore we will concentrate primarily on model compounds. Readers are encouraged to refer to the Butler chapter and other recent reviews for a more detailed coverage of VHPOs [134–136].

Haloperoxidases (HPOs) catalyze the 2-electron oxidation of halides ($X^- = Cl^-$, Br^-, or I^-) by hydrogen peroxide (Eq. 4). Haloperoxidases are referred to as chloroperoxidases (ClPOs), bromoperoxidases (BrPOs), or iodoperoxidases (IPOs) depending on the most electronegative halogen they are capable of oxidizing. The oxidation potentials of the halides are pH-dependent, and more acidic conditions are usually required to oxidize the more electronegative halides.

$$H_2O_2 + X^- + H^+ \xrightarrow{\text{HPO}} H_2O + OH^- + \text{"}X^+\text{"} \ (X^+ = XOH, X_2, X_3^-, \text{Enz-X})$$

$$(4)$$

Prior to 1984, the only known HPO enzymes contained a heme prosthetic group that catalyzes halide oxidation via an iron-centered redox mechanism. A vanadium-containing haloperoxidase of approximately 95 kDa [134] was first isolated from the brown alga *Ascophyllum nodosum* and the vanadium shown to be essential for catalytic activity [137, 138]. Vanadium haloperoxidases (VHPOs) have since been found in a wide variety of marine algae, seaweed, and a lichen [134]. Some of the halogenated natural products of these organisms have potential applications as antifungal, antibacterial, and antiviral agents and have even been implicated in the formation of the ozone hole over Antarctica [134, 139].

VHPOs are acidic proteins with similar sequence homology about the single vanadium(V) binding site [135]. Crystallographic data for VHPOs include a

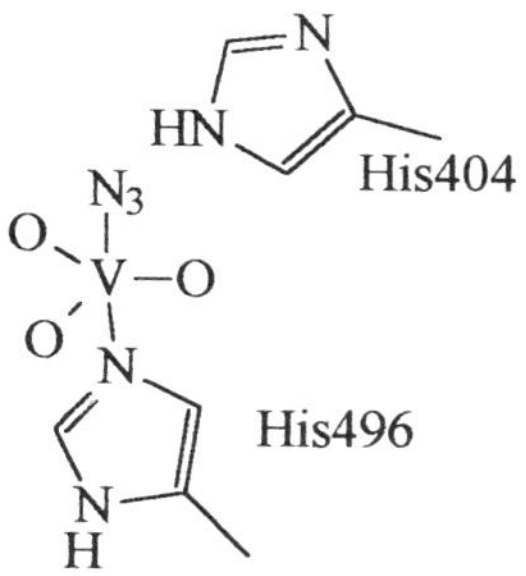

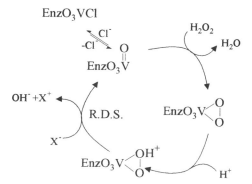

Fig. 11. Coordination environment of vanadium in the azide-bound form of VClPO from *C. inaequalis*

recent structure of an inactive azide-bound form of VClPO from the fungus *Curvularia inaequalis* and preliminary data on an azide-free form of the same enzyme [140]. In both structures, histidine is the only amino acid coordinated to the vanadium. Either three non-protein oxygens and azide (azide-bound form) or four non-protein oxygens (azide-free form) are bound to the vanadium to form a 5-coordinate trigonal bipyramidal center (Fig. 11). Histidine 404 may act as a crucial acid-base site ($pK_a > 5$), regulating both the rate of peroxide binding and the rate of subsequent halide oxidation (vide infra).

VHPOs are believed to catalyze the 2-electron oxidation of a halide via a Lewis acid-promoted mechanism. A redox mechanism involving initial oxidation of the metal center is not possible with VHPOs because the vanadium is already in its highest oxidation state. XAS [141, 142] and EPR [143] data suggest that vanadium remains in the +5 oxidation state through the entire catalytic cycle. The reaction is believed to proceed by a sequential order mechanism, with H_2O_2 adding initially. Both H_2O_2 and halide can compete for the site on or near the vanadium. Halide oxidation is thought to occur by Nucleophilic attack of the halide on the bound peroxide. Based on these enzymological studies and model compound reactivity described in section 7.2.3, a mechanism can be drawn for the reaction (Fig. 12) [144–147]. The rates of bromide and chloride

Fig. 12. Proposed mechanism for the VHPO-catalyzed oxidation of halides by hydrogen peroxide

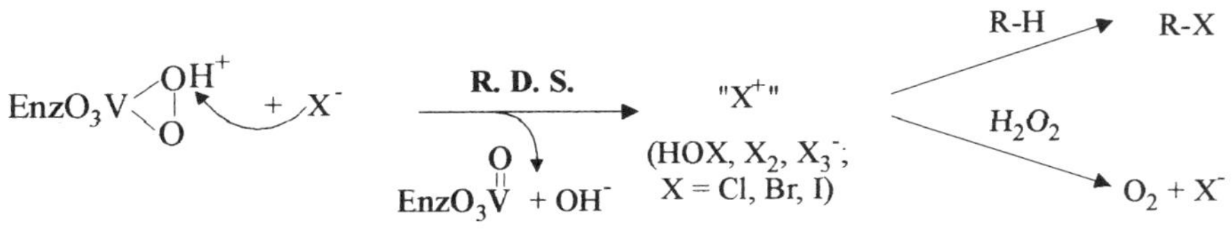

Fig. 13. Proposed mechanism for the formation of halogenated substrates and singlet oxygen involving nucleophilic attack of the halide on the peroxovanadate active site

oxidation are highly pH-dependent, with several proposed protonation equilibria thought to affect both ligand coordination and enzyme activation [144–147]. Protonation of histidine 404 is probably especially important for activation of the peroxovanadate complex, since model studies (vide infra) have shown that protonation of peroxovanadate complexes is essential for catalytic activity [148]. The protonated histidine 404 could activate the complex through hydrogen bonding interactions to the coordinated peroxide [140, 148].

The oxidized halide formed in the catalytic cycle is referred to as "X^+", but it probably takes the form of HOX, X_2, or X_3^-, depending on the type of halide, the pH, and the halide concentration in solution [146]. Kinetic studies support nucleophilic attack of the halide on the peroxovanadate active site as the rate-determining step (Fig. 13) [149]. There is some evidence that the oxidized halide may initially be enzyme-bound and only released if an organic substrate is not immediately available for halogenation [150]. Depending on the reaction conditions, the X^+ can either halogenate an organic substrate (enzyme-bound or not) or react with an additional equivalent of hydrogen peroxide to give singlet oxygen . The formation of dioxygen is favored in the absence of organic substrate and at higher pH. Steady state kinetic measurements for substrate halogenation and catalase activity agree within a factor of 2, implying that the rate-limiting step is the same for both processes.

7.2
Model Compounds

7.2.1
51V NMR Models

A wide variety of vanadium (V) compounds with O, N, S, and halide donor atoms have been characterized by ^{51}V NMR showing chemical shifts ranging from +1400 ppm (VS_4^{3-}) to −830 ppm ($VO(O_2)_3^{3-}$) vs $VOCl_3$ (Table 6). With the quantity of species now characterized, enough is known about trends in chemical shifts to (1) predict chemical shifts for known compounds and use ^{51}V NMR as an additional characterization technique and (2) predict approximate coordination environments for complexes where characterization is limited. ^{51}V chemical shifts can be qualitatively predicted by considering the electronegativity and hardness of the donor atom, and geometric strain caused by small chelate ring size or bulky substituents. Complexes with oxygen and nitrogen do-

Table 6. Representative ^{51}V NMR Shifts vs. $VOCl_3$ [47, 49]

Complex	δ (ppm)	Complex	δ (ppm)
VS_4^{3-}	1395	$VO(O_2)_3)^{3-}$	< -830
VOS_3^{3-}	740	$HVO(O_2)_3^{2-}$	-737.2
$VO_2S_2^{3-}$	184	$V(O_2)_4^{3-}$	-737.6
VO_3S^{3-}	-250	$VOBr_3$(neat)	432
VO_4^{3-}	-541.2	$VOCl_3$(neat)	0
HVO_4^{2-}	-538.8	VOF_3(THF)	-757 to -760
$H_2VO_4^-$	-560.4	VO(HSHED)(CAT)	221
VO_2^+	-546	VO(HSHED)(DBC)	382
$VO_3(O_2)^{3-}$	-769.3	VO(Br-HSHED)(CAT)	283
$HVO_3(O_2)^{2-}$	-627.6	VO(Br-HSHED)(DBC)	426
$VO(O_2)^+$	-549	VO(SALIMH)(CAT)	480
$VO_2(O_2)_2^{3-}$	-769.3	VO(SALIMH)(DBC)	600
$H_2VO_2(O_2)_2^-$	-699.9	VO(Br-SALIMH)(CAT)	513
$VO(O_2)_2(NH_3)^-$	-746	VO(Br-SALIMH)(DBC)	604

nors, such as water, hydroxide, alcohols, monodentate carboxylates, and amines, give shifts in the -400 to -600 ppm range. Shifts upfield of -600 ppm tend to result from negatively charged multidentate ligands that form 3- or 4-member chelate rings such as peroxides, η^2-CO_2^- and η^2-NO_3^- [151, 152]. The observed trends no longer hold when non-innocent ligands with low energy ligand-to-metal charge-transfer bands are used, such as catecholates or hydroxamates. These complexes are highly deshielded giving chemical shifts that are unusually far downfield (Table 6) [47].

Early ^{51}V NMR studies on VBrPO from *A. nodosum* show an unprecedented -1200 ppm chemical shift with a 3–6.5 kHz linewidth [153]. Based on this chemical shift value, the authors predicted a coordination environment containing 6 or 7 highly electronegative donors (O/N) with at least 2 η^2-carboxylate donors. Taking into consideration the subsequent data for model complexes, the coordination environment in the recent crystal structure of an inactive form of VClPO from *C. inaequalis*, and NMR data on vanadium-bound human apotransferrin, it is necessary to reconsider the validity of the peak at -1200 ppm. In the crystal structure of VClPO, the vanadium active site has a trigonal bipyramidal coordination with 3 non-peptide oxygen donors (i.e. O^{2-}, HO^-, or H_2O), an azide, and a histidine bound [140]. Assuming the azide displaces one of the oxygen donors, the active form of the enzyme would contain $H_xVO_4(his)^{x-3}$ which should give a peak slightly upfield of $H_xVO_4^{x-3}$ (~ -540 to -560 ppm). In vanadium-bound human apotransferrin, an 80 kDa protein, two vanadium binding sites can be distinguished upon titration with vanadate by the growth of two peaks at -529.5 and -531.5 ppm ($W_{1/2} \sim 400$ Hz) [154]. If the vanadium is bound very tightly in VBrPO, as it is in human apotransferrin, linewidths closer to those seen for transferrin would be expected. If the vanadium is more weakly bound, the peak may be broadened beyond detection levels.

82   C. Slebodnick · B. J. Hamstra · V. L. Pecoraro

7.2.2
ESEEM Models

While a histidine imidazole is the only protein-derived ligand bound to vanadium in the crystal structure of the VClPO [140], very few vanadium(IV) and
-(V) complexes with an imidazole ligand have been crystallographically
characterized. Most crystallographically characterized complexes of this type
contain the ligand HSALIMH (Fig. 14) [155, 156]. Only one of these complexes
is a vanadium(V) complex, stabilized in the higher oxidation state by a catecholate ligand [156].

The ESEEM spectrum of the reduced form of VBrPO is also consistent with
the presence of at least one imidazole ligand in the equatorial plane of the vanadyl ion (Fig. 15) [157]. Recent ESEEM studies of several model complexes
(Fig. 16) indicate that the additional low-frequency components of the ESEEM
spectrum of VBrPO may result from a nitrogen donor bound in an axial position with respect to the vanadyl ion [158]. This suggests that the inactivation of

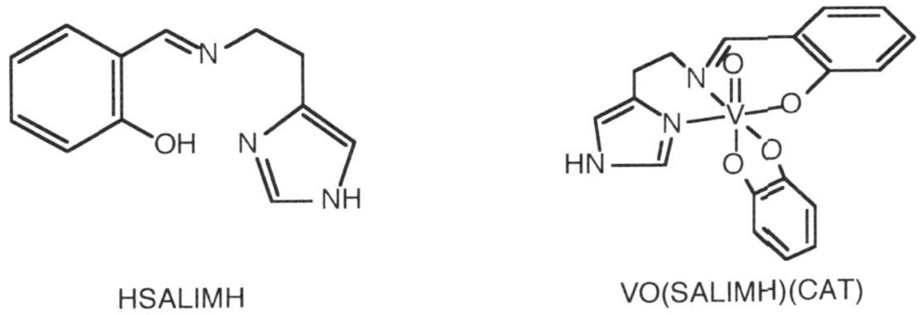

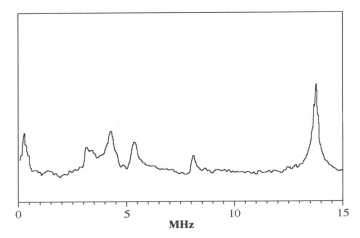

HSALIMH   VO(SALIMH)(CAT)

Fig. 14. Structures of the imidazole-containing ligand HSALIMH and its vanadium(V)-cate-
chol complex

Fig. 15. ESEEM spectrum of VBrPO from *A. nodosum* as measured by de Boer et al. (1988)
FEBS Letters 235:93

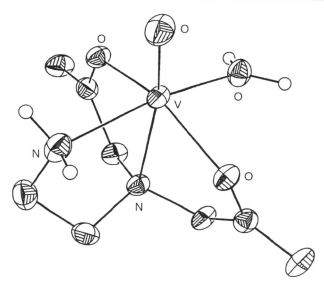

Fig. 16. ORTEP diagram of [VO(H$_2$O)aeida], a spectroscopic model for the effects of axial nitrogen ligation on ESEEM spectra of vanadyl complexes

reduced VBrPO may be caused by vanadium(IV) binding both the imidazole ligand, which would be used as an acid-base catalyst in the vanadium(V) form of the enzyme, and the originally bound imidazole (His-496 in ClPO; His-168 in BrPO from *A. nodosum*) [140]. As mentioned above, a nearby imidazole (His-404) has been shown to be present in the crystal structure of the ClPO, and sequence homology indicates that the bromoperoxidases also have suitable histidine residues near the active site which may perform the same function (His-101 in the VBrPO from *A. nodusum*, among others) [140].

7.2.3
Functional Models

The first mechanistic data for a functional mimic of V-HPO were obtained for *cis*-dioxovanadium (VO$_2^+$) in acidic aqueous solution [159, 160]. Both VHPO and VO$_2^+$ catalyze the stoichiometric formation of "Br$^+$" from hydrogen peroxide and bromide (Eq. 4). The oxidized bromide then proceeds to brominate an organic substrate or oxidize hydrogen peroxide to oxygen. However, unlike the VHPOs, which show optimal activity at pH 5–7, catalytic activity with VO$_2^+$ is only observed under highly acidic conditions (pH ≤ 2). The kinetics of bromide oxidation by VO$_2^+$ are consistent with a first-order rate dependence on bromide, but a second-order rate dependence on the vanadium concentration (Eq. 5). Clearly, this is unlike the enzymes where there is only one vanadium per sub-unit [143]. The observed rate is also a function of acid and peroxide concentrations. Based on a series of equilibrium and kinetic measurements, the mechanism in Fig. 17 was proposed. The active peroxovanadate species is V$_2$O$_2$(O$_2$)$_3$,

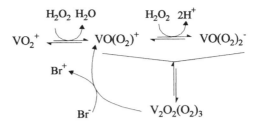

Fig. 17. Proposed mechanism for the oxidation of bromide by $V_2O_2(O_2)_3$

a minor species in solution formed from the dimerization of $VO(O_2)^+$ and $VO(O_2)_2^-$ (Eq. 6) according to the equilibrium: $K = V_2O_2(O_2)_3]/([VO(O_2)^+][VO(O_2)_2^-]$. The rate dependence on acid and peroxide concentrations results from the equilibria between $VO(O_2)^+$, $VO(O_2)_2^-$, and $V_2O_2(O_2)_3$, with a maximum rate occurring at relative acid and peroxide concentrations where $VO(O_2)^+$ and $VO(O_2)_2^-$ have equal concentrations and the concentration of $V_2O_2(O_2)_3$ is maximized. The highly acidic conditions are believed to be necessary for reactivity, because $VO(O_2)^+$ and $V_2O_2(O_2)_3$ only exist under these conditions [160]. Table 7 summarizes the second order rate constants for VBrPO and model systems.

$$- d[H_2O_2]/dt = d[Br^+]/dt = k[V_2O_2(O_2)_3][Br^-] = kK[VO(O_2)^+][VO(O_2)_2^-][Br^-] \tag{5}$$

$$VO(O_2)^+ + VO(O_2)_2^- \Leftrightarrow V_2O_2(O_2)_3 \tag{6}$$

As mentioned above, the highly acidic conditions under which vanadate catalyzes the oxidation of halides and the second-order dependence of the reactions on the vanadium concentration are in contrast with the slightly acidic conditions and first-order vanadium dependence under which VHPOs catalyze halide oxidation. Several complexes which act as functional models for VHPOs have recently been reported which address these issues.

The first reported monomeric vanadium complexes shown to act as functional models for VHPOs were the iminophenolate complexes reported by Butler et al. and illustrated in Fig. 18 [161]. The reaction of these complexes with

Table 7. Second order rate constants for bromide oxidation

Peroxovanadate Species	solvent	k ($M^{-1}s^{-1}$)	ref
V-BrPO (*A. nodosum*), pH 4	H_2O	1.75×10^5	[146]
V-BrPO (*A. nodosum*), pH 7.9	H_2O	2800	[146]
$V_2O_2(O_2)_3$	H_2O or 25% EtOH/H_2O	4.1 ± 0.1	[160]
$VO(O_2)(Hnta)^-$	CH_3CN	170 ± 30	[148]
$VO(O_2)(Hheida)^-$	CH_3CN	280 ± 40	[148]
$VO(O_2)(ada)^-$	CH_3CN	220 ± 30	[148]
$VO(O_2)(Hbpg)^+$	CH_3CN	21 ± 3	[148]
$VO(O_2)(tpa)^+$	CH_3CN	100 ± 30	[148]
H_2O_2	CH_3CN	3.7 ± 0.9	[148]

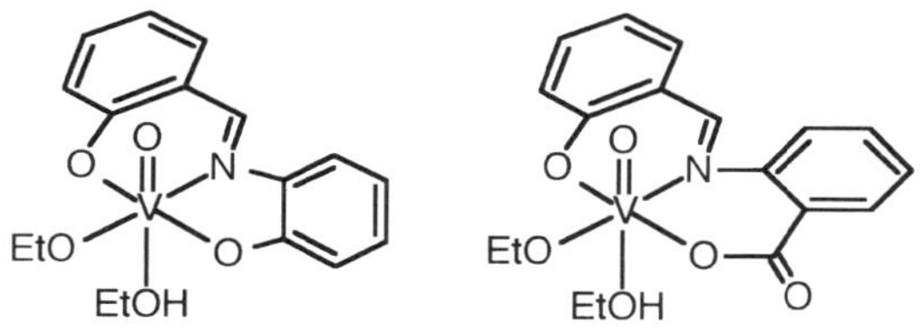

Fig. 18. Structures of iminophenolate complexes used as functional models for VHPOs

hydrogen peroxide, bromide, and 1,3,5-trimethoxybenzene resulted in the formation of 2-bromo-1,3,5-trimethoxybenzene over the course of 45 minutes to $1\,^1/_2$ hours, and was shown to be catalytic in the presence of added acid, carrying out up to 8 turnovers in the presence of sufficient peroxide and acid equivalents.

While a complete kinetic analysis of this reaction was not carried out due to bromination of the phenolate ligands, several experiments were performed which suggested a reasonable mechanism for halogenation. ^{51}V NMR and UV/visible spectra showed the partial conversion of the starting material to a monoperoxovanadium complex, and the conversion was shown to be complete within one hour under conditions in which the dioxovanadium complex was the sole vanadium species present in solution. Since the initial complex was not shown to be reactive with halide, this peroxovanadium species is probably the active oxidant in this system.

By examining the product formed by the bromination of 2,3-dimethoxytoluene, it was established that bromination of organic substrates in this system proceeds exclusively via an electrophilic mechanism, and that no bromine radical species were involved in the reaction (Fig. 19). From this work, the question of whether bromide coordinated directly to vanadium or directly attacked the bound peroxide of the peroxovanadium complex could not be determined.

Butler et al. also reported that the addition of hydroxide to a reaction mixture prevented bromination from occurring, while each additional equivalent of acid added enabled another turnover to take place. As ^{51}V NMR studies showed that OH$^-$ could not effectively compete with O_2^{2-} for coordination to vanadium, the role of acid in the proposed mechanism (Fig. 20) was believed to be in neutralizing the hydroxide formed in the reaction (Eq. 4).

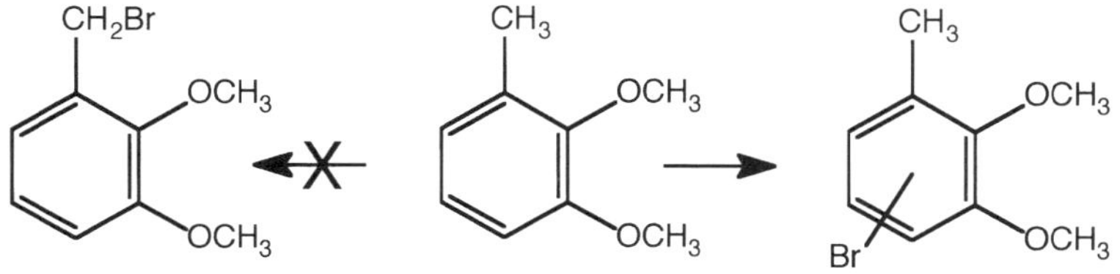

Fig. 19. Possible bromination reactions of 2,3-dimethoxytoluene. The pathway on the left corresponds to the product expected from a radical mechanism, and the pathway on the right corresponds to the product(s) expected from an electrophilic mechanism

Subsequent to this work, Pecoraro and co-workers reported that potassium salts of the peroxovanadium complexes of the ligands H_3heida, H_3nta, and H_2pmida (Fig. 21), when dissolved in acetonitrile (using 18-crown-6 as a solubilizing agent), rapidly and stoichiometrically oxidized iodide and bromide to triiodide and tribromide upon the addition of one equivalent of acid [162]. Ligands of this type were chosen in order to completely satisfy the coordination

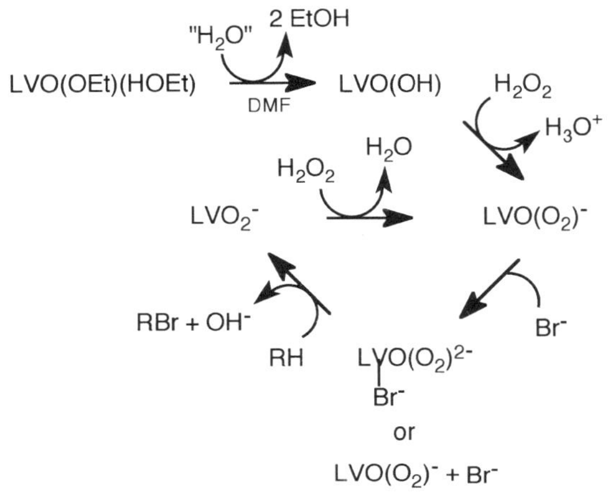

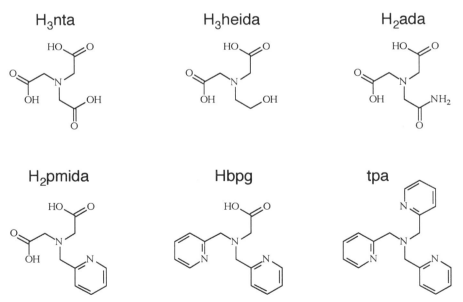

Fig. 20. Mechanism proposed by Butler and co-workers for the catalytic bromination of organic substrates by the complexes shown in Fig. 18

Fig. 21. Tetradentate ligands used in the synthesis of peroxovanadium complexes by Pecoraro et al.

sphere of vanadium when a single peroxide was bound in the equatorial plane. Fig. 22 shows the structure of [VO(O$_2$)Hheida]$^-$, which is the most efficient VHPO model known to date.

Unlike the previous studies by Butler's group, no reactivity (stoichiometric or catalytic) was observed in the absence of added acid. Bromination of organic substrates was shown spectrophotometrically by monitoring the conversion of Phenol Red to Bromophenol Blue (Fig. 23). The addition of excess equivalents of acid and peroxide resulted in up to 10 turnovers within 3 minutes, representing a rate of reaction at least an order of magnitude greater than that observed in any previous studies. Furthermore, the addition of hydroxide and peroxide to a solution of the oxidized halide in the absence of Phenol Red rapidly generated dioxygen gas, as observed for the VHPOs at higher pH values. Thus this

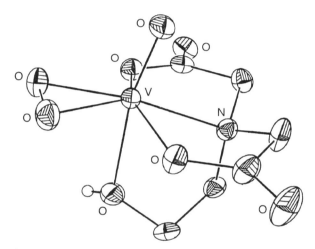

Fig. 22. ORTEP diagram of [VO(O$_2$)Hheida]$^-$

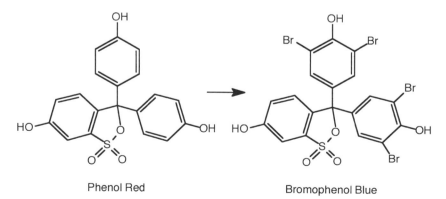

Phenol Red   Bromophenol Blue

Fig. 23. Structure of Phenol Red and its bromination product, Bromophenol Blue system provides the first non-vanadate model compounds which demonstrate both the halogenation and "catalase" reactions performed by the VHPOs.

Kinetic studies of these complexes revealed a first-order dependence on vanadium for all complexes studied, and at concentrations up to 2.5 mM (Fig. 24) for the complex $VO(O_2)bpg$ (Fig. 25), which exhibited the highest solubility in acetonitrile of any of the complexes studied, indicating that the ligands successfully prevented the formation of dimers as observed in the oxidation of bromide by vanadate [148].

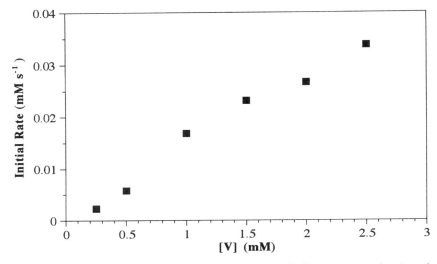

Fig. 24. Initial rate dependence of bromide oxidation on $VO(O_2)bpg$ concentration. Reaction conditions: 50 mM tetrabutylammonium bromide, 50 mM tetraethylammonium perchlorate, 1 equivalent of $HClO_4$ added per vanadium complex (CH_3CN solvent)

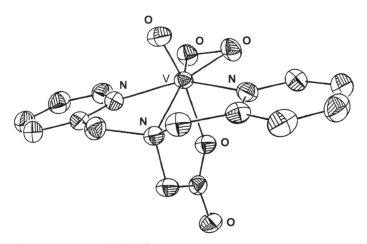

Fig. 25. ORTEP diagram of $VO(O_2)bpg$

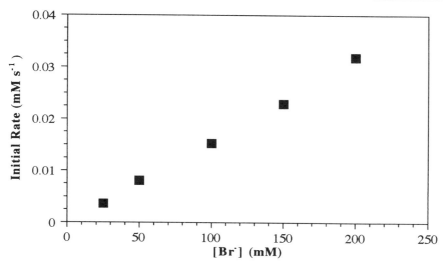

Fig. 26. Initial rate dependence of bromide oxidation on bromide concentration. Reaction conditions: 0.5 mM $VO(O_2)$bpg, 0.5 mM $HClO_4$ (CH_3CN solvent)

No spectroscopic evidence for coordination of bromide or iodide to vanadium was found, as no changes in the UV/visible spectra of the complexes were observed upon addition of excess bromide or iodide. Furthermore, a first-order dependence on the halide concentration was observed, with no saturation behavior up to 250 mM bromide (a 500-fold excess), supporting the hypothesis that halide coordination to vanadium is not part of the oxidation mechanism (Fig. 26).

In contrast to the inability of halides to produce changes in the spectra of the peroxovanadium complexes, the addition of one equivalent of acid in the absence of halides produced small shifts in the peroxo-to-vanadium charge-transfer bands of these complexes, which reverted to their original positions upon subsequent addition of one equivalent of base (Fig. 27) [148]. Kinetic studies of the proton dependence of the halide reaction clearly showed saturation behavior (Fig. 28), and data at different acid concentrations were fitted according to a mechanism in which a protonation pre-equilibrium preceded the halide oxidation step (taking into account the buffering ability of the halides in CH_3CN), yielding pK_a values for the complexes and second-order rate constants for the halide oxidation reactions (Table 7). The second-order rate constants thus obtained for the complexes are between 5 to 70 times larger than those reported for the vanadate dimer in aqueous solution at pH values ≤ 2.

The oxygen-generating reactions of these complexes were also examined, although kinetic studies were not reported. The addition of $H_2^{18}O_2$ to a solution of oxidized bromide produced dioxygen gas which was exclusively ^{18}O-labeled, indicating that the peroxide is oxidized without oxygen-oxygen bond cleavage.

Based on these experiments, Pecoraro et al. proposed the mechanism illustrated in Fig. 29, which encompasses both the halogenation and "catalase" reac-

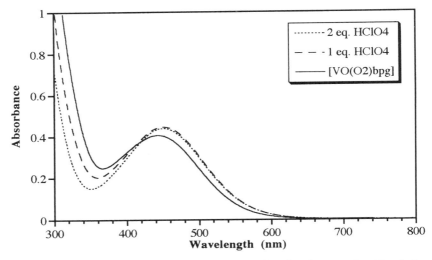

Fig. 27. Spectroscopic observation of the titration of $HClO_4$ into a 1.0 mM solution of $VO(O_2)bpg$ (CH_3CN solvent)

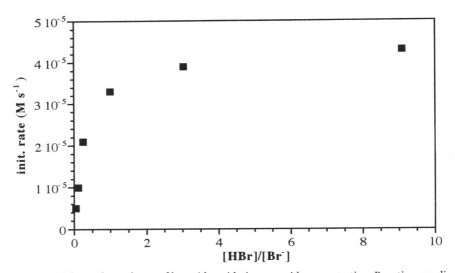

Fig. 28. Initial rate dependence of bromide oxidation on acid concentration. Reaction conditions: 0.5 mM $[VO(O_2)nta]^{2-}$, 25 mM tetrabutylammonium bromide, 25 mM tetraethylammonium perchlorate (CH_3CN solvent)

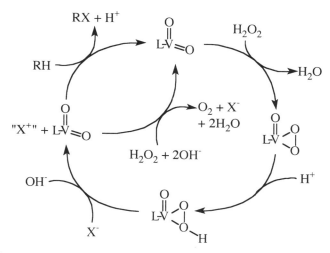

Fig. 29. Mechanism proposed by Pecoraro and co-workers of the catalytic oxidation of halides, and the halogenation of organic substrates or generation of dioxygen by the oxidized halogen species thus formed

tions of the model complexes and, by extension, the VHPOs. In this mechanism, halides are oxidized by peroxovanadium complexes via nucleophilic attack by the halide on a hydroperoxovanadium complex rather than by direct coordination of the halide to vanadium. The designation of the site of protonation of the vanadium complex is not directly supported by any spectroscopic evidence, however, it is chemically reasonable when compared to the proposed mechanism for the acid-catalyzed reaction of hydrogen peroxide with halides, in which an $H_3O_2^+$ intermediate has been proposed [163]. A side-on bound hydroperoxo ligand bound to a vanadium(V) center may be considered to be functionally equivalent to $H_3O_2^+$. Furthermore, a side-on bound alkylperoxovanadium complex has been characterized and shown to be an oxidation catalyst, lending additional support to the proposal of a side-on bound hydroperoxo ligand [164]. This mechanism does not directly specify the nature of the active oxidized halogen species, as the trihalide is the only product observed under conditions of excess halide, but the formation of hypohalous acid is (at least initially) indicated, as HOX is formally equivalent to $OH^- + X^+$ (where X = chloride, bromide, or iodide).

It should be noted that the mechanism proposed by Pecoraro and co-workers is also consistent with the data reported previously by Butler's group. Although at first glance the necessity for protonation of the complex shown by Pecoraro et al. appears to be at variance with the proton-independent capacity to stoichiometrically oxidize bromide reported by Butler and co-workers, in reality the binding of peroxide in Butler's complexes results in the liberation of one equivalent of acid, which could then protonate the bound peroxide in a subsequent step prior to the oxidation of the halide.

As mentioned above, the acid-base catalyst in this mechanism that is necessary for the generation of the hydroperoxovanadium complex in the enzymes is likely to be a histidine residue, which may have several functions in the catalytic cycle. In addition to the necessity for protonation of the peroxide in generating the active oxidant at the appropriate time in the catalytic cycle, proton transfer to an oxo or hydroxo ligand from the imidazole may be essential for labilizing such ligands for replacement by a peroxo ligand or ligands, whose binding ability may be enhanced by deprotonation by a histidine imidazole. Furthermore, the inactivity of the reduced form of the enzyme may be due to coordination of the essential histidine imidazole, which may either inhibit the reoxidation of vanadium(IV) to vanadium(V), or remain bound to the reoxidized vanadium center and therefore be incapable of acting as an acid-base catalyst.

Chloride oxidation has been observed for several of the model systems, but to the best of our knowledge, no rate constants for chloride oxidation have been reported because the reactions are too slow to measure. Manipulation of solvent polarity by controlling the water concentration in acetonitrile led to chloride oxidation by one of several possible vanadium species and a preliminary lower limit for the rate constant based on total vandium concentration was obtained [165]. Chloride oxidation rates have been determined by monitoring the rate of conversion of cresol red to chlorocresol purple (Fig. 30). [K(18-C-6)$_2$][VO$_3$] dissolved in water/acetonitrile solutions (2–25% water by volume) in the presence of acid (1–10 equiv.) and hydrogen peroxide (1.5 or more equivalents) yields V$_2$O$_2$(O$_2$)$_3$ as the primary species (Eq. 6) according to the ^{51}V NMR [165].

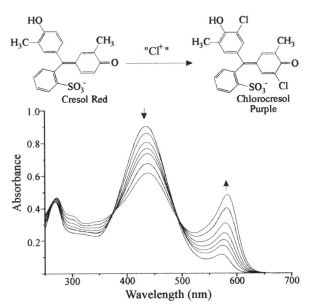

Fig. 30. Sample UV/vis spectrum for monitoring the rate of conversion of Cresol Red to Chlorocresol Purple. Reaction conditions: 0.5 mM VO$_3^-$, 10 mM H$_2$O$_2$, 10 mM HClO$_4$, 25 mM Cl$^-$, and 1 mM Cresol Red in 6% water/94% acetonitrile

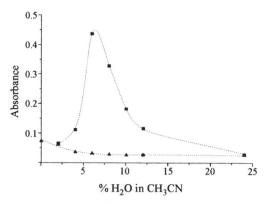

Fig. 31. Rate dependence of chloride oxidation on water concentration. Reaction conditions: 10 mM H_2O_2, 10 mM $HClO_4$, 2 mM VO_3^-, t = 2 hrs (squares); 10 mM H_2O_2, 10 mM $HClO_4$, t = 4 hrs (triangles)

The rate of chloride oxidation is strongly dependent on the water concentration, and it appears likely that higher water concentrations are required at higher chloride concentrations. Under the chloride concentrations in most studies to date (25–50 mM Cl⁻) 6% water in acetonitrile is the optimum condition (Fig. 31). The rate of chloride oxidation shows a first-order dependence on vanadium concentration (Fig. 32a) giving a rate constant of 1.3×10^{-3} min⁻¹, which corresponds to a half-life of 8.8 h. Chloride acts as a substrate and an inhibitor (Fig. 32b), similar to VClPO [147]. Although no rate constants are currently available for bromide oxidation under acetonitrile/water conditions, the rates are clearly faster than those observed in 0–25% ethanol in water for $V_2O_2(O_2)_3$.

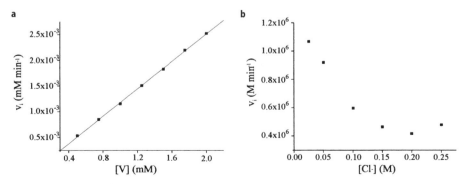

Fig. 32. Initial rate of chlorocresol purple formation as a function of **a** vanadium concentration and **b** chloride concentration. Reaction conditions: **a** 10 mM H_2O_2, 10 mM $HClO_4$, 25 mM Et_4NCl, 0.5 mM Cresol Red in 6% water/94% acetonitrile; **b** 10 mM H_2O_2, 10 mM $HClO_4$, 1 mV VO_3, 1 mM Cresol Red in 6% water/94% acetonitrile

The rate dependence on solvent conditions appears to be more than just a function of driving the equilibrium in Eq. 6 to the right. Presumably the less polar solvent conditions activate the complexes by increasing the pK_a of the catalytic compound(s), enabling the formation of an active protonated species. There is also the possibility that changing the solvent conditions varies the degree of chloride inhibition, thus affecting the rate.

7.3
Vanadium Peroxidases

VHPOs may actually be vanadium peroxidases (VPOs) responsible for introduction of a variety of functional groups into marine natural products, rather than just halogens. The discovery and classification as a HPO may have arisen because of the halogen-rich marine environment in which most VHPO-containing organisms live. However, based on the currently proposed mechanism for VHPO activity, electron-rich substrates that can access the vanadium active site should be targets for electrophilic oxidation. It has already been shown that VHPO from *A. nodosum* can oxidize cyanide [166] and thiocyanate [167], which are both functional groups found in some marine natural products. Evidence for VHPO oxidation of sulfide would be especially interesting since VHPO and vanadium-bound PPTP have very similar structures at the vanadium active site. The primary difference is coordination of a histidine to the vanadium in VHPO and a cysteine in PPTP. If VHPO is able to oxidize sulfides, it would provide additional evidence for the proposed mechanism of peroxovanadate insulin mimicry involving oxidation of the vanadium-bound cysteine to sulfoxide.

Additional evidence for the potential peroxidase activity of VHPO is provided by model peroxovanadates. The variety of reactions that have been catalyzed by peroxovanadates is summarized in Fig. 33. They were reviewed recently by Butler and co-workers [42]. Alkene, sulfide, and phosphine oxidations are all catalyzed by electrophilic attack of the peroxovanadate to yield epoxide, sulfoxide, or phosphine oxide, respectively. Peroxovanadates also catalyze oxidation of aromatics, alkanes, and sulfoxides, but in these cases, the mechanism is proposed to involve a V(IV) radical species.

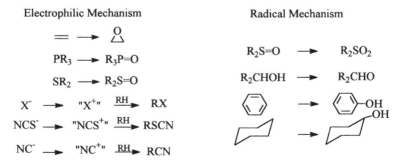

Fig. 33. Reactions catalyzed by peroxovanadium compounds

8
Nitrogenase

8.1
Nitrogenase

Several different types of nitrogen-fixing bacteria are known, each of which contains nitrogenase enzymes that catalyze the reduction of atmospheric N_2 to NH_3. For many years molybdenum was believed to be essential for nitrogenase activity, but within the past 15 years it has become apparent that nitrogen fixation can also occur in Mo-deficient environments via the expression of vanadium-dependent nitrogenases. It should be emphasized that V-dependent nitrogenase activity does not arise simply by the substitution of vanadium for molybdenum in the more commonly found Mo-dependent enzymes; the vanadium nitrogenases are distinct enzymes, although they appear to be structurally and functionally very similar to the Mo-dependent nitrogenases. For the purposes of this review, only the functional differences between the Mo- and V-dependent enzymes and the known structural information which relates to the coordination environment of vanadium will be discussed in detail; extensive reviews of vanadium nitrogenases are recommended for a thorough treatment of the subject [168, 169].

The reduction of N_2 to NH_3 by vanadium-dependent nitrogenases can be described by the following equation:

$$N_2 + 12e^- + 12H^+ + 24MgATP \rightarrow 2NH_3 + 3H_2 + 24MgADP + 24PO_4^{3-}$$

$$(7)$$

This activity is markedly less than that observed for the corresponding molybdenum enzymes, which produce only one mole of H_2 per 2 moles of N_2 (and consequently require one-third less ATP, protons, and reducing equivalents) [170]. However, the ability of the V-dependent enzymes to maintain higher activity at lower temperatures than the Mo-dependent enzymes may compensate in part for this relative inefficiency [171].

Vanadium-dependent nitrogenases also naturally produce small amounts of products which are not produced by the Mo-dependent enzymes under normal conditions. Small amounts of N_2H_4 are observed during the reduction of N_2, and the alternative substrate C_2H_2 produces small amounts of C_2H_6 in addition to the C_2H_4 normally observed as the product of acetylene reduction by nitrogenases [172, 173].

Apart from these differences, the chemistry of the vanadium- and molybdenum-dependent nitrogenases is quite similar, and information obtained from the study of one of the two types of enzymes is generally applicable to the study of the other. The similarities between the two enzymes appear to extend to structural considerations as well.

Like the molybdenum nitrogenases, vanadium nitrogenases consist of two distinct proteins. The first is an iron-containing homodimeric protein which contains an $[Fe_4S_4]$ cluster and a nucleotide-binding site. This protein is believed to act as an ATP-dependent electron donor to the second protein,

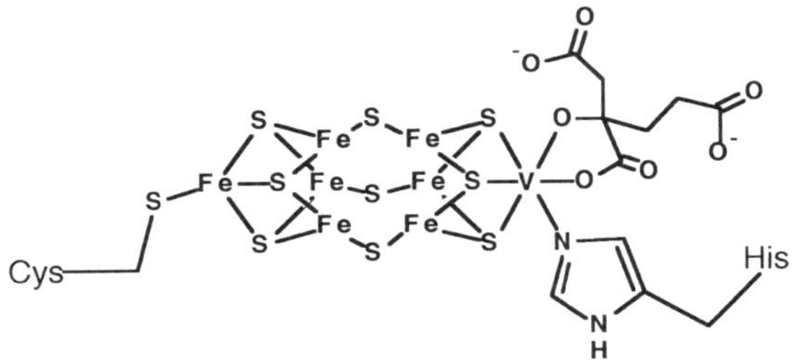

Fig. 34. Probable structure of the vanadium-containing cofactor in the vanadium-dependent nitrogenases

which has a $N\alpha_2\beta_2\delta_2$ subunit structure. This second protein contains four unusual low-potential [Fe_8S_8] clusters (P-clusters) which may have an electron-transfer role and two [VFeS] clusters (M-clusters) to which substrates are believed to bind and be reduced. (The third type of subunit in the vanadium-containing protein is not observed in the corresponding molybdenum-containing protein.)

Analysis of the isolated VFeS cluster showed that the ratio of atoms in the cluster was approximately 1:6:5[174]. X-ray absorption spectroscopy and EXAFS data suggest that vanadium is in an oxidation state between +2 and +4, that its coordination environment is distorted octahedral, and that it is probably in a cuboidal-type cluster environment similar to the [VFe_3S_4] cluster synthesized and characterized by Kovacs and Holm [175–178]. EPR studies of the vanadium-containing protein are consistent with an $S = 3/2$ ground state for the vanadium-containing cluster [179]. MCD spectroscopic data are also consistent with this spin state assignment [180].

Given these data, which are all quite similar to those obtained for the corresponding MoFeS clusters in the molybdenum-dependent nitrogenases, and the other similarities between the vanadium- and molybdenum-dependent enzymes, it is not unreasonable to assume that the VFeS cluster has the structure shown in Fig. 34 (or one quite similar–there are at least some minor environmental differences), which is based on the recently determined structures of the MoFeS clusters in molybdenum nitrogenases [181, 182]. This structure still awaits crystallographic confirmation.

8.2
Model Compounds

The probable structure of the vanadium-containing cofactor in nitrogenases offers several challenges with respect to the chemical synthesis of the cofactor. To date no synthesis of the cluster found in the molybdenum- or vanadium-dependent nitrogenases has been reported. However, several clusters have been synthesized in the last ten years which show some structural similarities to the cofactors of these enzymes.

Much of the work in this area is based on the cluster [$VFe_3S_4Cl_3(DMF)_3$]$^-$ (Fig. 35), first reported by Kovacs and Holm in 1986 [178]. Subsequent studies

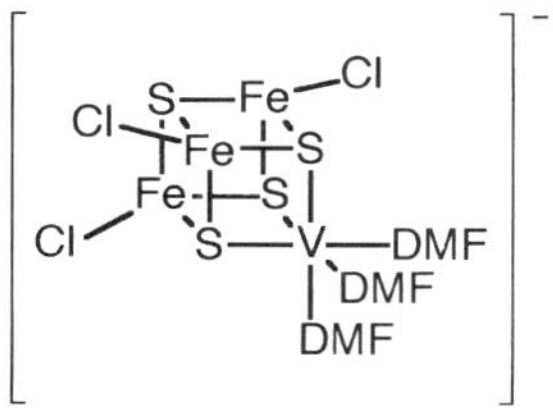

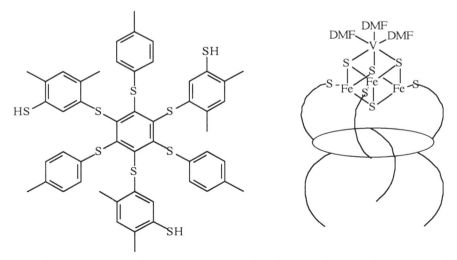

Fig. 35. Structure of $[VFe_3S_4Cl_3(DMF)_3]^-$

of this compound revealed functional, structural, and spectroscopic similarities between VFe_3S_4 and $MoFe_3S_4$ clusters [183–185], and established clusters of this type as excellent structural models of the vanadium coordination environment in nitrogenase [177].

Regiospecific examination of the ligand binding ability of the vanadium site in cubanes of this type was accomplished through the use of the cavitand ligand $L(SH)_3$ (Fig. 36) to selectively bind to the three iron atoms in the cubane cluster and leave the vanadium site free for ligand substitution reactions [186]. Azide was shown to bind weakly and incompletely to vanadium when present in excess, and cyanide was also found to bind somewhat strongly. Cyanide binding to vanadium led to a –0.5 V shift in the reversible oxidation potential for the VFe_3S_4 cluster, suggesting that ligand binding may have a significant effect on the reactivity of the analogous cluster in nitrogenase.

Perhaps the synthetic compound which most closely resembles the structure of the VFeS cluster to date is $VFe_4S_6(PEt_3)_4Cl$ (Fig. 37), which contains 11 of the 17 atoms believed to be present in the VFeS cofactor (based on analogy to the

Fig. 36. Structure of the cavitand ligand $L(SH)_3$ and a schematic of its complex with a VFeS cubane cluster

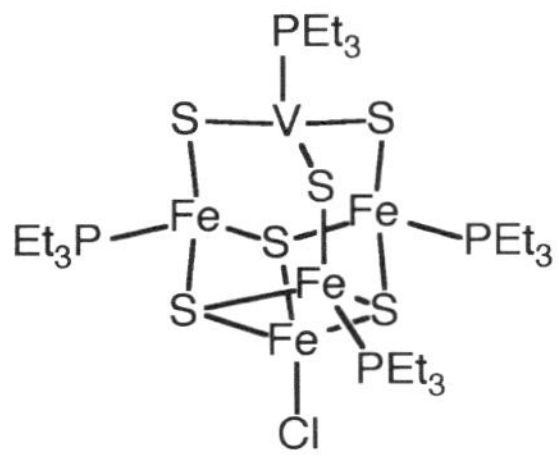

Fig. 37. Structure of $VFe_4S_6(PEt)_3Cl$

Mo-containing enzyme), and 10 of these 11 atoms are arranged with the proper connectivity, though they do not necessarily possess the correct coordination number when compared to their counterparts in the enzyme [187].

Still, for some time no catalytic activity was observed for clusters of this type until the recent report of the catalytic conversion of hydrazine to ammonia by $[VFe_3S_4Cl_3(DMF)_3]^-$ and related cubanes by Coucouvanis and co-workers [188]. Hydrazine is known to be an alternative substrate for the nitrogenases, and is also a potential intermediate in the reduction of dinitrogen [172]. By reacting $[VFe_3S_4Cl_3(DMF)_3]^-$ with cobaltocene (a reducing agent), 2,6-lutidinium hydrochloride (a proton source with a non-coordinating base), and N_2H_4 in CH_3CN (a relatively non-coordinating solvent), multiple turnovers were observed within a 2-hour period.

While no kinetic data were reported for this system, several mechanistic experiments suggest that an available coordination site at vanadium, and not iron, is important for mediating the catalytic reduction of N_2H_4 in these systems. Figure 38 shows that clusters in which the iron-bound chlorides were exchanged for bromide or iodide, while possessing similar redox potentials and having the same number of available coordination sites, show little if any difference in the rate of reaction with N_2H_4. Conversely, clusters in which one or more of the DMF solvent molecules coordinated to vanadium were exchanged for less labile ligands show significantly inhibited reactivity (Fig. 39), as do reactions performed in DMF solution. Based on these and other results, Coucouvanis et al. proposed the scheme shown in Fig. 40 for the reduction of N_2H_4 by cubanes of this type.

While coordination to vanadium appears to be critical in the reduction of hydrazine by cuboidal clusters of this type, it remains unclear from these studies whether or not coordination to iron is important in the early stages of dinitrogen reduction by nitrogenases (particularly in light of the apparent coordinatively unsaturated iron sites in these enzymes). Dinitrogen reduction by VFeS (and MoFeS) clusters remains an unrealized goal in synthetic systems.

Several non-cubane vanadium complexes are known to coordinate and activate dinitrogen. While these complexes are structurally quite different from the active site cluster of nitrogenase, considerably more success has been achieved in modeling the reactivity of nitrogenase in these systems.

The dinitrogen-bridged complexes $[(V(C_6H_4CH_2N(CH_3)_2)_2(pyridine))_2N_2]$ and $[Na(V(C_6H_2(CH_3)_3)_3)_2N_2]^-$ (Fig. 41) react with protic solvents forming am-

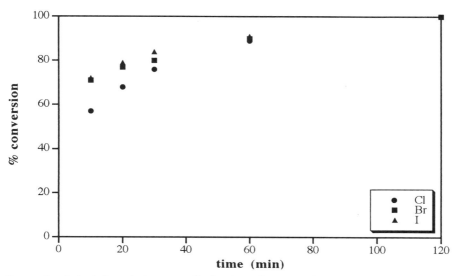

Fig. 38. Graph depicting relative rates of hydrazine reduction with chloro-, bromo-, and iodo-ligated VFe_3S_4 clusters. From data reported by Malinak et al. (1995) J Am Chem Soc 117:3126

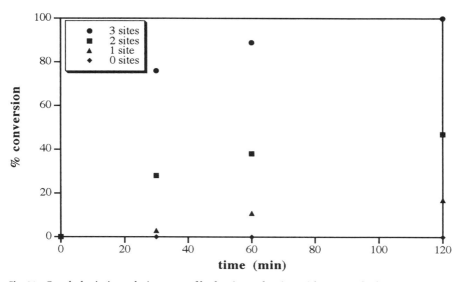

Fig. 39. Graph depicting relative rates of hydrazine reduction with $VFe_3S_4Cl_3$ clusters in which varying numbers of labile coordination sites on vanadium are available. 3 sites: 3 exchangeable solvent molecules bound to V. 2 sites: 1 triethylphosphine & 2 exchangeable solvent molecules bound to V. 1 site: 2,2'-bipyridine and 1 exchangeable solvent molecule bound to V. 0 sites: tris(pyrazolyl)borate bound to V. From data reported by Malinak et al. (1995, J Am Chem Soc 117:3126)

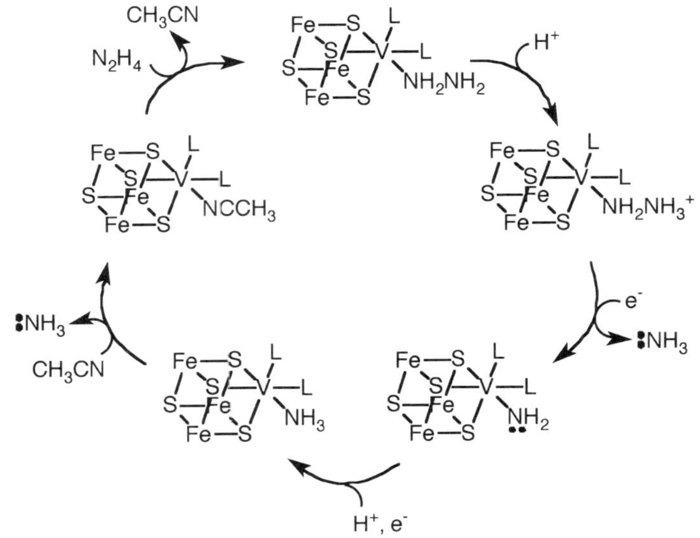

Fig. 40. Mechanism proposed by Coucouvanis and co-workers for the catalytic reduction of hydrazine by VFeS cubanes

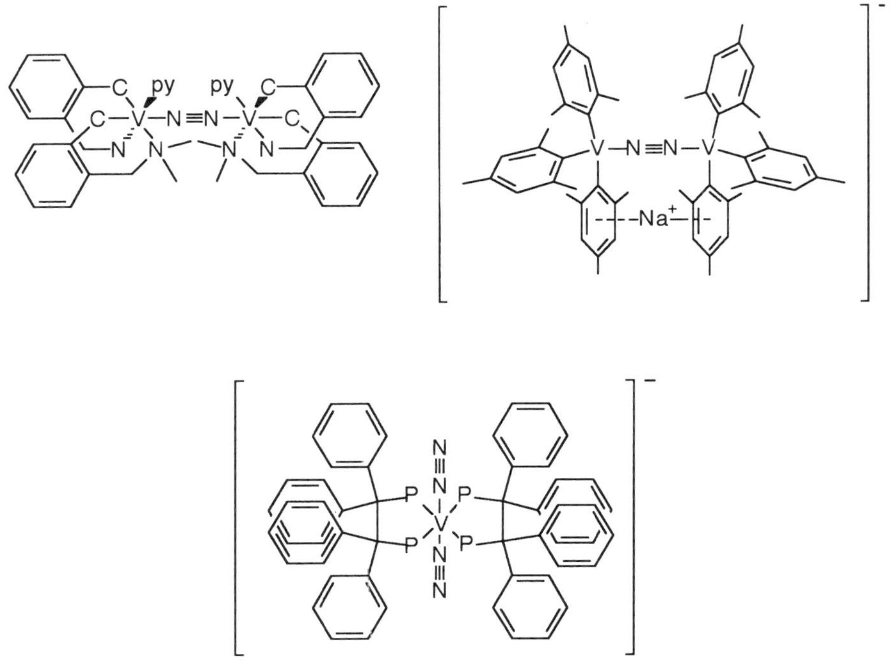

Fig. 41. Non-cubane functional models for vanadium nitrogenase

monia, and may therefore serve as models for bimetallic binding and activation of dinitrogen, which may take place at the active site of nitrogenase [189, 190]. The mononuclear complex $[V(dppe)_2(N_2)_2]^-$ (Fig. 41) which contains end-on bound dinitrogen, generates ammonia and small amounts of hydrazine upon reaction with acid [191], and therefore may be a model for the monometallic binding and activation of dinitrogen, which, as shown in studies by Coucouvanis et al. (vide supra), may also be the mode of action of nitrogenase. In none of these non-cubane systems, however, has catalytic activity been reported, and so, as with the synthesis of the vanadium-containing cluster of nitrogenase, the synthesis and characterization of vanadium complexes that catalytically reduce dinitrogen remains an elusive goal.

9
Conclusions

Many advances have been made over the past decade in elucidating the role and mechanism of action of vanadium in biological systems. Vanadium functions as an essential element in a number of biological systems including tunicates, *A. muscaria* mushrooms, and vanadium haloperoxidase- and vanadium nitrogenase-containing organisms. In the vanadium-accumulating tunicates and mushrooms, the role of vanadium is essentially unknown with current theories concentrating primarily on defense mechanisms. Vanadium haloperoxidases are believed to generate a variety of potentially useful antifungal and antibacterial halogenated products as well as harmful halogenated pollutants in the marine environment. Thus, one of the primary roles for VHPOs may also be defense, with the unfortunate production of other less directed toxins. A mechanism for VHPO activity has been proposed and is strongly supported by studies of model compounds. The active site structure and several mechanisms for nitrogen reduction by vanadium nitrogenases have also been proposed, based, to a large extent, on studies of model compounds. Vanadium's ability to bind to and inhibit phosphate-related proteins, affect the reactivity of PPTPs, ATPases, and possibly other phosphate-related enzymes has led to applications as insulin mimics and in protein characterization by photocleavage.

Given the similarity of vanadate to phosphate, we would like to end by considering the relationship between VHPOs and phosphatases. A number of crystal structures have recently been reported of phosphatases with vanadate bound as a transition state analog [195-197]. Although these proteins show little sequence and structural similarity throughout the protein backbone, the vanadium binding sites are very similar among the phosphatases and compared to VClPO (Fig. 42). The vanadium is trigonal bipyramidal with a single axial bound histidine (VClPO and rat acid phosphatase) or cysteine (PPTP). Arginine is conserved at the active sites and appears to be important in stabilizing the trigonal bipyramidal geometry by forming one or more strong hydrogen bonds with the vanadate oxygens through both side chain and amide nitrogens. Histidine and/or aspartic acid are also located nearby, presumably functioning as acid-base catalysts. The acid phosphatase is especially similar to VClPO; both enzymes have a hydrophobic region near the active site, which for VClPO is be-

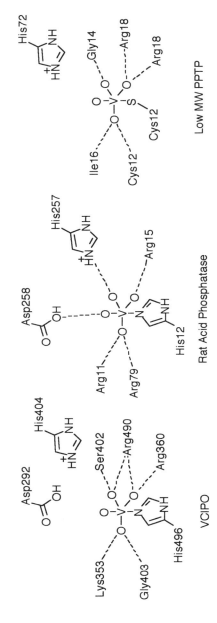

Fig. 42. Schematics of the active sites of three potentially related enzymes with vanadate bound: VClPO [140], rat acid phosphatase [196], and low molecular weight PPTP [195]

lieved to stabilize chloride binding. In addition, a crystal structure of rat acid phosphatase with chloride bound near the active site has been reported, proving halide accessibility [198].

Based on the obvious similarities between these structures, the possibility that VHPOs are actually phosphatases should be explored. If the enzymes do prove to exhibit phosphatase activity, one must consider the possibility that VHPOs function in the cell as phosphatases. The isolation of a vanadate bound form may be due to the relatively high concentrations of vanadium and halide in sea water. Another intriguing possibility is that these enzymes represent an evolutionary change from phosphatase to haloperoxidase, providing these organisms with a function that they need for survival, but can not obtain due to an absence of heme-containing HPOs.

Recognition of the relationship between phosphatases and VHPO leads to an even more profound prediction when one considers the corollary of the above discussion. That is, can phosphatases isolated from organisms that are not exposed to high levels of vanadium naturally, such as humans, be converted to VHPOs in the presence of vanadate? Our laboratory is presently investigating this fascinating question. If such an activity is detected, the results could have potential importance in the study of insulin mimicry. To date, there are two general proposals for the mechanism of vanadium action as an insulin mimic: (1) vanadate temporarily deactivates PPTP by inhibiting the enzyme at the phosphate binding site and (2) peroxovanadates cause more permanent damage by selectively oxidizing the cysteine that functions as a nucleophile during the tyrosine dephosphorylation mechanism. If the binding of peroxovanadates at the phosphate binding site in phosphatases initiates a new function, haloperoxidation, a third possibility is that non PPTP sites are halogenated or peroxidated resulting in activation and/or deactivation of other, currently unsuspected enzymes. Furthermore, one must consider potential new toxicities that could arise by converting a phosphatase to a hypochlorite or hypobromite generator within cells.

Note added in proof: Ref. 200 which appeared in March 1997 reports that the apo VClPO exhibits phosphatase aktivity as predicted above.

Acknowledgments: Financial support from the National Institutes of Health (Grant GM42703) is gratefully acknowledged. C.S. thanks the National Institutes of Health for a postdoctoral fellowship (1F32GM18370-01). We would also like to thank Dr. Gerard Colpas and Prof. Charles Root for useful discussions.

10
References

1. Chasteen ND (1990) Vanadium in Biological Systems. Kluwer, Dordrecht, The Netherlands
2. Sigel H, Sigel A (1995) Metal Ions in Biological Systems 31. Marcel Dekker, New York
3. Hammond CR (1990) The Elements. In: Weast RC (ed) CRC Handbook of Chemistry and Physics. CRC, Boca Raton, p B-5
4. Boas LV, Pessoa JC (1987) In: Wilkinson G (ed) Comprehensive Coordination Chemistry, The Synthesis, Reactions, Properties and Applications of Coordination Compounds. Pergamon Press, Oxford, p 453

5. Bonadies JA, Butler WM, Pecoraro VL, Carrano CJ (1987) Inorg Chem 26:1218
6. Meier R, Boddin M, Mitzenheim S, Kanamori K (1995) Met Ions Biol Sys 31:45
7. Bond MR, Czernuszewicz RS, Dave BC, Yan Q, Mohan M, Verastegue R, Carrano CJ (1995) Inorg Chem 34:5857
8. Kanamori K, Teraoka M, Maeda H, Okamoto K (1993) Chem Lett 1731
9. Garner CD, Collison D, Mabbs FE (1995) Met Ions Biol Sys 31:617
10. Pasquali M, Marchetti F, Floriani C (1979) Inorg Chem 18: 401
11. Carrondo MAAFdCT, Duarte MTLS, da Silva JJRF, da Silva JAL (1992) Struct Chem 3:113
12. Cooper SR, Koh YB, Raymond KN (1982) J Am Chem Soc 104:5092
13. Pasquali M, Marchetti F, Floriani C, Cesari M (1980) Inorg Chem 19:1198
14. Nesterova YM, Anan'eva NN, Polynova TN, Porai-Koshits MA, Pechurova NI (1977) Doklady Akademii Nauk SSSR 2:350
15. Pasquali M, Marchetti F, Floriani C, Merlino S (1977) J Chem Soc, Dalton Trans 139
16. Ballhausen CJ, Gray HB (1962) Inorg Chem 1:111
17. Wüthrich K (1965) Helv Chem Acta 48:1012
18. Cornman C, R., Zovinka EP, Boyajian YD, Geiser-Bush KM, Boyle PD, Singh P (1995) Inorg Chem 34:4213
19. Chasteen ND (1981) In: Berliner LJ , Reuben J (eds) Biological Magnetic Resonance. Plenum, New York, p 53
20. Eaton SS, Eaton GR (1990) Biological Applications of EPR, ENDOR, and ESEEM Spectroscopy. In: Chasteen ND (ed) Vanadium in Biological Systems. Kluwer, Boston, p 199
21. Rehder D (1991) Angew Chem Int Ed Eng 30:148
22. Rehder D (1990) Biological Applications of ^{51}V NMR Spectroscopy. In: Chasteen ND (ed) Vanadium in Biological Systems. Kluwer Academic Publishers, Boston, p 173
23. Rehder D (1995) Met Ions Biol Sys 31:1
24. Kustin K, Robinson WE (1995) Met Ions Biol Sys 31:511
25. Gresser MJ, Tracey AS (1990) In: Chasteen ND (ed) Vanadium in Biological Systems. Kluwer, Dordrecht, p 63
26. Stankiewicz PJ, Tracey AS, Crans DC (1995) Met Ions Biol Sys 31:287
27. Scheidt WR, Tsai C, Hoard JL (1971) J Am Chem Soc 93:3867
28. Scheidt WR, Collins DM, Hoard JL (1971) J Am Chem Soc 93:3873
29. Scheidt WR, Countryman R, Hoard JL (1971) J Am Chem Soc 93:3878
30. Giacomelli A, Floriani C, Duarte AOdS, Chiesi-Villa A, Guastini C (1982) Inorg Chem 21:3310
31. Drueckhammer DG, Durrwachter JR, Pederson RL, Crans DC, Daniels L, Wong C-H (1989) J Org Chem 54:70
32. Crans DC, Simone CM, Blanchard JS (1992) J Am Chem Soc 114:4926
33. Lindquist RN, Lynn J, Leinhard GE (1973) J Am Chem Soc 95:8762
34. Wlodawer A, Miller M, Sjölin L (1983) Proc Natl Acad Sci USA 80:3628
35. Crans DC, Chen H, Anderson OP, Miller MM (1993) J Am Chem Soc 115:6769
36. Colpas GJ, Hamstra BH, Kampf JW, Pecoraro VL (1994) Inorg Chem 33:4669
37. Nanda KK, Sinn E, Addison AW (1996) Inorg Chem 35:1
38. Plass W (1996) Inorg Chim Acta 244:221
39. Pecoraro V, Bonadies JA, Marrese CA, Carrano CJ (1984) J Am Chem Soc 106:3360
40. Riley PE, Pecoraro VL, Carrano CJ, Bonadies JA, Raymond KN (1986) Inorg Chem 25:154
41. Armstrong EM, Beddoes RL, Calviou LJ, Charnock JM, Collison D, Ertok N, Naismith JH, Garner CD (1993) J Am Chem Soc 115:807
42. Butler A, Clague MJ, Meister GE (1994) Chem Rev
43. Shaver A, Ng JB, Hall DA, Lum BS, Posner BI (1993) Inorg Chem 32:3109
44. Wu D-X, Lei X-J, Cao R, Hong M-C (1992) Jiegou Huaxue 11:65
45. Crans DC, Shin PK (1994) J Am Chem Soc 116:1305
46. Crans DC, Ehde PM, Shin PK, Pettersson L (1991) J Am Chem Soc 113:3728
47. Cornman CR, Colpas GJ, Hoeschele JD, Kampf J, Pecoraro VL (1992) J Am Chem Soc 114:9925

48. Crans DC, Shin PK (1988) Inorg Chem 27:1797
49. Howarth OW (1990) Prog Nucl Magn Reson Spectrosc 22:453
50. Crans DC, Shin PK, Armstrong KB (1995) Application of NMR Spectroscopy to Studies of Aqueous Coordination Chemistry of Vanadium(V) Complexes. In: Thorp HH , Pecoraro VL (eds) Mechanistic Bioinorganic Chemistry. American Chemical Society, Washington, D.C., p 303
51. Henze M (1911) Hoppe-Seyler's Z Physiol Chem 72:494
52. Henze M (1912) Hoppe-Seyler's Z Physiol Chem 79:215
53. Henze M (1913) Hoppe-Seyler's Z Physiol Chem 86:340
54. Ishii T, Nakai I, Okoshi K (1995) Met Ions Biol Sys 31:491
55. Hawkins CJ, Klott P, Parry DL, Swinehart JH (1983) Comp Biochem Physiol B:Comp Biochem 76B:555
56. Smith M, J., Dooseop K, Horenstein B, Nakanishi K (1991) Acc Chem Res 24:117
57. Kustin K, Robinson WE, Smith MJ (1990) Inverterbr Reprod Dev 17:129
58. Smith MJ, Ryan DE, Nakanishi K, Frank P, Hodgson KO (1995) Met Ions Biol Sys 31:423
59. Michibata H, Sakurai H (1990) Vanadium in Ascidians. In: Chasteen ND (ed) Vanadium in Biological Systems. Kluwer, Boston, p 153
60. Anderson DH, Berg JR, Swinehart JH (1991) Comp Biochem Physiol A: Comp Physiol 99A:151
61. Anderson DH, Swinehart JH (1991) Comp Biochem Physiol A: Comp Physiol 99A:585
62. Wright RK (1981) Urochordates. In: Ratcliffe NA , Rowley AF (eds) Invertebrate Blood Cells. Academic Press, London, p 565
63. Oltz EM, Bruening RC, Smith MJ, Kustin K, Nakanishi K (1988) J Am Chem Soc 110:6162
64. Botte L, Scippa S, de Vincentiis M (1979) Experientia 35:1228
65. Oltz EM, Pollack S, Delohery T, Smith MJ, Ojika M, Lee S, Kustin K, Nakanishi K (1989) Experientia 45:186
66. Carlson RMK (1975) Proc Nat Acad Sci U S A 72:2217
67. Frank P, Carlson RMK, Hodgson KO (1988) Inorg Chem 27:118
68. Brand SG, Hawkins CJ, Parry DL (1987) Inorg Chem 26:627
69. Dingey AL, Kustin K, Macara IG, McLeod GC (1981) Biochim Biophys Acta 649:493
70. Frank P, Carlson RMK, Hodgson KO (1986) Inorg Chem 25:470
71. Taylor SW, Hawkins CJ, Parry DL, Swinehart JH, Hanson GR (1994) J Inorg Biochem 56:97
72. Lee S, Kustin K, Robinson WE, Frankel RB, Spartalian K (1988) J Inorg Biochem 33:183
73. Frank P, Kustin K, Robinson WE, Linebaugh L, Hodgson KO (1995) Inorg Chem 34:5942
74. Pirie BJS, Bell MV (1984) J Exp Mar Biol Ecol 74:187
75. Frank P, Hedman B, Carlson RMK, Hodgson KO (1994) Inorg Chem 33:3794
76. Hawkins CJ, James GA, Parry DL, Swinehart JH, Wood AL (1983) Comp Biochem Physiol B: Comp Biochem 76B:559
77. Agudelo MI, Kustin K, McLeod GC (1983) Comp Biochem Physiol A: Comp Physiol 75A:211
78. Michibata H, Iwata Y, Hirata J (1991) J Exp Zool 257:306
79. Bulls AR, Pippin CG, Hahn FE, Raymond KN (1990) J Am Chem Soc 112:2627
80. Ryan DE, Ghatlia ND, McDermott AE, Turro NJ, Nakanishi K (1992) J Am Chem Soc 114:9659
81. Ryan DE, Grant KB, Nakanishi K (1996) Biochemistry 35:8640
82. Lee S, Nakanishi K, Chiang MY, Frankel RB, Spartalian K (1988) J Chem Soc, Chem Commun 785
83. Ryan DE, Grant KB, Nakanishi K, Frank P, Hodgson K (1996) Biochemistry 35:8651
84. Smith MJ (1989) Experientia 45:452
85. ter Meulen H (1931) Recl Trav Chim Pays-Bas 50:491
86. Bayer E, Kneifel H (1972) Z Naturforsch 27b:207
87. Bayer E (1995) Metal Ions Biol Sys 31:407
88. Kneifel H, Bayer E (1973) Angew Chem Int Ed Engl 12:508
89. Kneifel H, Bayer E (1986) J Am Chem Soc 108:3075

90. Nawi MA, Riechel TL (1987) Inorg Chim Acta 136:33
91. Anderegg G, Koch E, Bayer E (1987) Inorg Chim Acta 127:183
92. Bayer E, Koch E, Anderegg G (1987) Angew Chem Int Ed Engl 26:545
93. Wieghardt K, Quilitzsch U, Nuber B, Weiss J (1978) Angew Chem Int Ed Engl 17:351
94. Guedes da Silva MFC, da Silva JAL, Fraústo da Silva JJR, Pombeiro AJL, Amatore C, Verpeaux J-N (1996) J Am Chem Soc 118:7568
95. Felcman J, Vaz MCTA, Fraústo da Silva JJR (1984) Inorg Chim Acta 93:101
96. Thackerey RD, Riechel TL (1988) J Electroanal Chem 245:131
97. Shechter Y (1990) Diabetes 39:1
98. Shechter Y, Meyerovitch J, Farfel Z, Sack J, Bruck R, Bar-Meir S, Amir S, Degani H, Karlish SJD (1990) Insulin Mimetic Effects of Vanadium. In: Chasteen ND (ed) Vanadium in Biological Systems. Kluwer, Boston, p 129
99. Orvig C, Thompson KH, Battell M, McNeill JH (1995) Met Ions Biol Sys 31:575
100. Hall DA (1996) Ph.D. thesis, McGill University
101. Tolman EL, Barris E, Burns M, Pansisni A, Partridge R (1979) Life Sci 25:1159
102. Domingo JL, Gomez M, Sanchez DJ, Llobet JM, Keen CL (1995) Mol Cell Biochem 153:233
103. Degani H, Gochin M, Karlish SJD, Schechter Y (1981) Biochemistry 20:5795
104. Cros GH, Cam M, C., Serrano J-J, Ribes G, McNeill JH (1995) Mol Cell Biochem 153:191
105. Sun Y, James BR, Rettig SJ, Orvig C (1996) Inorg Chem 35:1667
106. Caravan P, Gelmini L, Glover N, Herring G, Li H, McNeill JH, Rettig SJ, Setyawati IA, Shuter E, Sun Y, Tracey AS, Yuen VG, Orvig C (1995) J Am Chem Soc 117:12759
107. McNeill JH, Yuen VG, Dai S, Orvig C (1995) Mol Cell Biochem 153:175
108. Sakurai H, Tsuchiya K, Nukatsuka M, Kawada J, Ishikawa SI, Yoshida H, Komatsu M (1990) J Clin Biochem Nutrit 8:193
109. Kadota S, Fantus IG, Deragon G, Guyda HJ, Hersh B, Posner BI (1987) Biochem Biophys Res Commun 147:259
110. Kadota S, Fantus IG, Deragon G, Guyda HJ, Posner BI (1987) J Biol Chem 262:8252
111. Yale JF, Vigeant C, Nardolillo C, Chu Q, Yu J-Z, Shaver A, Posner BI (1995) Mol Cell Biochem 153:181
112. Bevan AP, Drake PG, Yale J-F, Shaver A, Posner BI (1995) Mol Cell Biochem 153:49
113. Shaver A, Ng JB, Hall DA, Posner BI (1995) Mol Cell Biochem 153:5
114. Shechter Y, Li J, Meyerovitch J, Gefel D, Bruck R, Elberg G, Miller DS, Shisheva A (1995) Mol Cell Biochem 153:39
115. Li J, Elberg G, Crans DC, Schechter Y (1996) Biochemistry 35:8314
116. Battell ML, Yuen VG, McNeill JH (1992) Pharmacol Commun 1:291
117. Goldfine AB, Simonson DC, Folli F, Patti M-E, Kahn CR (1995) Mol Cell Biochem 153:217
118. Lee-Eiford A, Ow RA, Gibbons IR (1986) J Biol Chem 261:2337
119. Gibbons IR, Mocz G (1990) Vanadate Sensitized Photocleavage of Proteins. In: Chasteen ND (ed) Vanadium in Biological Systems. Kluwer, Boston, p 143
120. Muhlrad A, Ringel I (1995) Met Ions Biol Sys 31:211
121. Stryer L (1988) Biochemistry, 3rd edn. W. H. Freeman, New York
122. Gibbons IR, Lee-Eiford A, Mocz G, Phillipson CA, Tang W-JY, Gibbons BH (1987) J Biol Chem 262:2780
123. Lobert S, Isern N, Hennington BS, Correia J (1994) Biochemistry 33:6244
124. Gibbons IR, Gibbons BH, Mocz G, Asai DJ (1991) Nature 352:640
125. Grammer JC, Cremo CR, Yount RG (1988) Biochemistry 27:8408
126. Cremo CR, Grammer JC, Yount RG (1988) Biochemistry 27:8415
127. Cremo CR, Grammer JC, Yount RG (1989) J Biol Chem 264:6608
128. Cremo CR, Loo JA, Edmonds CG, Hatlelid KM (1992) Biochemistry 31:491
129. Tang W-JY, Gibbons IR (1987) J Biol Chem 263:17728
130. Cremo CR, Long GT, Grammer JC (1990) Biochemistry 29:7982
131. Muhlrad A, Peyser M, Ringel I (1991) Biochemistry 30:958
132. Mocz G (1989) Eur J Biochem 179:373
133. Mocz G, Gibbons IR (1990) J Biol Chem 265:2917

134. Vilter H (1995) Met Ions Biol Sys 31:325
135. Butler A, Walker JV (1993) Chemical Reviews 93:1937
136. Wever R, Krenn BE (1990) Vanadium Haloperoxidases. In: Chasteen ND (ed) Vanadium in Biological Systems. Kluwer, Dordrecht, p 81
137. Vilter H (1984) Phytochem 23:1387
138. Wever R, Plat H, de Boer E (1985) Biochim Biophys Acta 830:181
139. Neidleman SL, Geigert JL (1986) Biohalogenation. Ellis Horwood, New York
140. Messerschmidt A, Wever R (1996) Proc Nat Acad Sci USA 93:392
141. Arber JM, de Boer E, Garner CD, Hasnain SS, Wever R (1989) Biochemistry 28:7968
142. Hormes J, Kuetgens U, Chauvistre R, Schriber W, Anders N, Vilter H, Rehder D, Weidemann C (1988) Biochim Biophys Acta 956:293
143. de Boer E, Koon K, Wever R (1988) Biochemistry 27:1629
144. Everett RR, Soedjak HS, Butler A (1990) J Biol Chem 265:15671
145. Soedjak HS, Butler A (1991) Biochim Biophys Acta 1079:1
146. de Boer E, Wever R (1988) J Biol Chem 263:12326
147. Van Schijndel JWPM, Barnett P, Roelse J, Vollenbroek EGM, Wever R (1994) Eur J Biochem 225:151
148. Colpas GJ, Hamstra BJ, Kampf JW, Pecoraro VL (1996) J Am Chem Soc 118:3469
149. Everett RR, Butler A (1989) Inorg Chem 28:393
150. Tschirret-Guth RA, Butler A (1994) J Am Chem Soc 116:411
151. Priebsch W, Rehder D (1985) Inorg Chem 24:3058
152. Rehder D, Weidemann C, Duch A, Priebsch W (1988) Inorg Chem 27:584
153. Vilter H, Rehder D (1987) Inorg Chim Acta 136:L7
154. Butler A, Eckert H (1989) J Am Chem Soc 111:2802
155. Cornman CR, Kampf J, Lah MS, Pecoraro VL (1992) Inorg Chem 31:2035
156. Cornman CR, Kampf J, Pecoraro VL (1992) Inorg Chem 31:1981
157. de Boer E, Keijzers CP, Klaassen AAK, Reijerse EJ, Collison D, Garner CD, Wever R (1988) FEBS Lett 235:93
158. Hamstra BJ, LoBrutto R, Houseman ALP, Colpas GJ, Kampf JW, Frasch WD, Pecoraro VL (unpublished results)
159. de la Rosa RI, Clague MJ, Butler A (1992) J Am Chem Soc 114:760
160. Clague MJ, Butler A (1995) J Am Chem Soc 117:3475
161. Clague MJ, Keder NL, Butler A (1993) Inorg Chem 32:4754
162. Colpas GJ, Hamstra BH, Kampf JW, Pecoraro VL (1994) J Am Chem Soc 116:3627
163. Bakac A, Assink B, Espenson JH, Wang W-D (1996) Inorg Chem 35:788
164. Mimoun H, Chaumette P, Mignard M, Saussine L, Fischer J, Weiss R (1983) Nouv J Chem 7:467
165. Slebodnick C, Colpas GJ, Pecoraro VL (unpublished results)
166. Wever R, Kustin K (1990) Adv Inorg Chem 35:81
167. Walker JV, Butler A (1996) Inorg Chim Acta 243:201
168. Eady RR (1990) Vanadium Nitrogenases. In: Chasteen ND (ed) Vanadium in Biological Systems. Kluwer, Boston, p 99
169. Eady RR (1995) Met Ions Biol Sys 31:363
170. Dilworth MJ, Eldridge ME, Eady RR (1993) Biochem J 289:359
171. Miller RW, Eady RR (1988) Biochem J 256:429
172. Dilworth MJ, Eady RR (1991) Biochem J 277:465
173. Dilworth MJ, Eady RR, Robson RL, Miller RW (1987) Nature 327:167
174. Smith BE, Eady RR, Lowe DJ, Gormal C (1988) Biochem J 250:299
175. Arber JM, Dobson BR, Eady RR, Stevens P, Hasnain SS, Garner CD, Smith BE (1987) Nature 325:327
176. George GN, Coyle CL, Hales BJ, Cramer SP (1988) J Am Chem Soc 110:4057
177. Chen J, Christiansen J, Tittsworth RC, Hales BJ, George SJ, Coucouvanis D, Cramer SP (1993) J Am Chem Soc 115:5509
178. Kovacs JA, Holm RH (1986) J Am Chem Soc 108:340
179. Hales BJ, True AE, Hoffman BM (1989) J Am Chem Soc 111:8519

180. Morningstar JE, Johnson MK, Case EE, Hales BJ (1987) Biochemistry 26:1795
181. Bolin JT, Ronco AE, Mortenson LE, Morgan TV, Xuong NH (1993) Proc Natl Acad Sci USA 90:1078
182. Chan MK, Kim J, Rees DC (1993) Science 260:792
183. Kovacs JA, Holm RH (1987) Inorg Chem 26:702
184. Kovacs JA, Holm RH (1987) Inorg Chem 26:711
185. Carney MJ, Kovacs JA, Zhang Y-P, Papaefthymiou GC, Spartalian K, Frankel RB, Holm RH (1987) Inorg Chem 26:719
186. Ciurli S, Holm RH (1989) Inorg Chem 28:1685
187. Norlander E, Lee SC, Cen W, Wu ZY, Natoli CR, Di Cicco A, Fillipponi A, Hedman B, Hodgson KO, Holm RH (1993) J Am Chem Soc 115:5549
188. Malinak SM, Demadis KD, Coucouvanis D (1995) J Am Chem Soc 117:3126
189. Leigh GJ, Prieto-Alcón A, Sanders J (1991) J Chem Soc, Chem Commun 921
190. Ferguson VR, Solari E, Floriani C, Chiesi-Villa A, Rizzoli C (1993) Angew Chem Int Ed Engl 32:396
191. Rehder D, Woitha C, Priebsch W, Gailus H (1992) J Chem Soc, Chem Commun 364
192. Michibata H, Hirata J, Uesaka M, Numakunai T, Sakurai H (1987) J Exp Zool 244:33
193. Michibata H, Uyama T, Hirata J (1990) Zool Sci 7:55
194. Saraste M, Sibbald PR, Wittinghofer A (1990) Trends Biochem Sci 15:430
195. Zhang M, Zhou M, Van Etten RL, Stauffacher CV (1997) Biochemistry 36:15
196. Lindqvist Y, Schneider G, Vihko P (1994) Eur J Biochem 221:139
197. Denu JM, Lohse DL, Vijayalakshmi J, Saper MA, Dixon JE (1996) Proc Nat Acad Sci, USA 93:2493
198. Schneider G, Lindqvist Y, Vihko P (1993) EMBO J 12:2609
199. Stryer L (1988) Biochemistry, 3rd edn. WH Freeman and Company New York
200. Hemrika W, Renirie R, Dekker HL, Barnett P, Wever R (1997) Proc Nat Acad Sci (USA) 94:2145

Vanadium Bromoperoxidase and Functional Mimics

Alison Butler* and Anne H. Baldwin

Department of Chemistry, University of California, Santa Barbara, CA 93106-9510
E-mail: butler@sbmm1.ucsb.edu

Haloperoxidases are enzymes which catalyze the oxidation of halides (iodide, bromide and chloride) by hydrogen peroxide, resulting in the halogenation of appropriate organic substrates or the halide-assisted disproportionation of hydrogen peroxide forming dioxygen. Two classes of vanadium haloperoxidases are known: vanadium bromoperoxidase, isolated mainly from marine algae, and vanadium chloroperoxidase from terrestrial fungi. A review of the structure, reactivity and enzyme kinetics of these enzymes is presented. Kinetic and mechanistic studies of several functional mimics of the vanadium haloperoxidases are also presented, including peroxo-complexes of vanadium(V) and other transition metal ions. These catalysts have proved useful in testing hypotheses concerning the role of vanadium in the haloperoxidase enzymes.

Key words: Vanadium, haloperoxidase, bromoperoxidase, chloroperoxidase, biomimic.

List of Abbreviations 110

1   Introduction 110

2   Structural Features of the Vanadium Haloperoxidases 112

3   Reactivity of the Vanadium Haloperoxidases 114
3.1 Halogenation and Halide-Assisted Disproportionation of
     Hydrogen Peroxide 114
3.2 Nature of the Oxidized Halogen Intermediate and
     the Question of Organic Substrate Binding to V-BrPO 115
3.3 The Steady-State Kinetics of V-ClPO and V-BrPO 116
3.4 Reversible and Irreversible Inactivation of V-BrPO
     Under Turnover Conditions 117
3.5 Reactivity of V-BrPO with Pseudohalides 118
3.6 Peroxide Reactivity of V-BrPO 119

4   Functional Mimics of Vanadium Haloperoxidase 119
4.1 cis-VO$_2^+$ in Acidic Aqueous Solution and Other Aqueous
     Vanadium(V) Species 119
4.2 Molybdenum(VI), Tungsten(VI) and Rhenium(VII) Complexes .. 122

* Corresponding Author.

Structure and Bonding, Vol. 89
© Springer Verlag Berlin Heidelberg 1997

4.3   The Hydroxyphenyl-Salicylideneamine Complex of VO(OH) 124
4.4   Other Complexes 126

5   On the Mechanism of V-HPO: Consideration of the Reaction
        Sequence, the Role of Vanadium and the Role of the Protein 127

6   References 130

List of Abbreviations

MCD   monochlorodimedone
V-BrPO   vanadium bromoperoxidase
V-HPO   vanadium haloperoxidase
V-ClPO   vanadium chloroperoxidase

1
Introduction

Haloperoxidases are enzymes which catalyze the oxidation of halides (iodide, bromide and chloride) by hydrogen peroxide, resulting in the halogenation of appropriate organic substrates [1] or the halide-assisted disproportionation of hydrogen peroxide forming dioxygen:

$$X^- + H_2O_2 + R{-}H + H^+ \rightarrow R{-}X + 2\,H_2O \tag{1}$$

$$X^- + 2\,H_2O_2 \rightarrow O_2 + 2\,H_2O + X^- \tag{2}$$

In the halogenation reaction, stoichiometric consumption of peroxide and protons occurs per equivalent of halogenated substrate produced. Fluoride is not a substrate because hydrogen peroxide does not have the potential to oxidize fluoride.

Two classes of haloperoxidase are known, the vanadium- and FeHeme-containing enzymes. Within the class of vanadium-haloperoxidase enzymes, both vanadium bromoperoxidases (V-BrPO), which are isolated mainly from marine algae, and vanadium chloroperoxidases (V-ClPO), which are isolated mainly from terrestrial fungi, have been identified. Many halogenated natural products have been isolated from marine organisms. These compounds range from volatile halogenated hydrocarbons, e. g., bromoform, chloroform, etc, which are produced in very large quantities [2–4], to chiral halogenated terpenes, acetogenins and indoles, among others, which are produced in smaller amounts, but which often have important biological activities (e.g., antifungal or antibacterial properties, feeding deterrents, etc) or pharmacological properties (e.g., antineoplastic activities) (Fig. 1). The reader is referred to a series of review articles on these natural products [5 and references therein]. Thus the function of the marine haloperoxidases is thought to be the biosynthesis of these natural products. Investigations of the role of the vanadium haloperoxidases and other marine haloperoxidases in the biosynthesis of such compounds are increasing.

The physiological role of vanadium chloroperoxidase (V-ClPO) remains to be determined. Halogenated natural products have not been identified in the

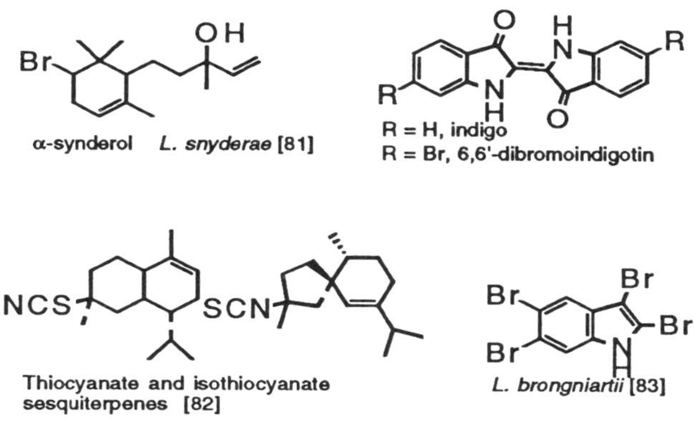

α-synderol   L. snyderae [81]

R = H, indigo
R = Br, 6,6'-dibromoindigotin

Thiocyanate and isothiocyanate
sesquiterpenes [82]

L. brongniartii [83]

Fig. 1

fungi which produce V-ClPO, including V-ClPO from *Curvularia inaequalis* (the most studied of the V-ClPOs) and other dematiaceous hyphomycetes [6]. V-ClPO is reported to be secreted from these fungi and, as will be discussed below, produces hypochlorous acid (HOCl). HOCl is a strong bactericidal agent which may be produced as a defense mechanism, or as an attack mechanism in the invasion of the plant cell wall of the fungi's host.

The vanadium bromoperoxidases are all acidic glycoproteins [7, 8] with a very similar amino acid composition [9], molecular weight, charge (pI 4–5), and vanadium content. The subunit molecular weight of V-BrPO is ca. 65,000. The isolated form of V-BrPO contains a sub-stoichiometric ratio of vanadium per subunit. However, a content of one gram-atom of vanadium/subunit can be achieved by addition of excess vanadate and subsequent removal of adventitiously bound vanadium(V) by dialysis [10–12]. Isozymes of vanadium bromoperoxidase which differ in carbohydrate content have been isolated from *A. nodosum* [7, 13]. While the physical characteristics of V-BrPO isolated from marine algae are all very similar, some differences in reactivity have been observed, such as specific activity [14].

V-BrPO (*A. nodosum*) is particularly stable to strong oxidants, such as singlet oxygen ($^1O_2$, $^1\Delta_g$) and oxidized bromine and chlorine compounds (HOBr, HOCl, etc) [13, 15]. V-BrPO also has appreciable thermal stability [16] and stability in many organic solvents. For example V-BrPO does not lose activity when stored at room temperature for a month in 60% (v/v) acetone, methanol or ethanol or 40% (v/v) 1-propanol [15]. In addition V-BrPO retains activity after immobilization on solid support media, e.g., photo-crosslinked to DEAE cellulofine [17].

V-ClPO (*C. inaequalis*) is a 67,488 Da protein consisting of 609 amino acid residues, as determined from the DNA sequence analysis [18]. This protein lacks disulfide bonds, although two cysteine residues are present as free thiols. Two putative *N*-glycosylation sites were identified, however the protein is not glycosylated. V-ClPO is secreted from *C. inaequalis*. The isolated form may have a variable vanadium content, depending on the concentration of vanadate in the growth medium. However, one vanadium(V) per subunit can be achieved by addition of excess vanadate to the growth medium or to the purified protein [19, 20] Like V-BrPO, V-ClPO is stable in the presence of organic substrates, at elevated temperatures and in the presence of high concentrations of strong oxidants (e.g., HOCl) [21].

Vanadium can be removed from V-HPO producing the inactive apo-enzyme derivative. The activity can be fully restored by addition of vanadate [19, 22]. Other metal ions have not been found to restore haloperoxidase activity [23]. Vanadium(V) is only fully incorporated in the absence of phosphate [6, 22, 24]. Like phosphate, molybdate, arsenate, tetrafluoroaluminate, and tetrafluoroberrylate inhibit coordination of vanadium(V) to apo-BrPO [25].

2
Structural Features of Vanadium Haloperoxidases

Recently, the crystal structure of vanadium chloroperoxidase isolated from *Curvularia inaequalis* was reported by Messerschmidt and Wever [26] (see the chapter by Messerschmidt in the succeeding volume). The X-ray structure (2.11 Å resolution) of the 67 kDa protein showed a cylindrical shape of approximately 80 Å by 55 Å. Analysis of the secondary structure revealed extended strands and loop regions and short β-strands arranged in pairs as anti-parallel β-ladders, but the main structural motif is α-helical, with two four- helix bundles. Vanadate is coordinated at the top of one of these bundles in a broad channel which is lined on one half with predominantly polar residues including an ion-pair between Arg-360 and Asp-292 and several main chain carbonyl oxygens (Fig. 2). The other half of the channel is hydrophobic, containing Pro-47, Pro-211, Try-350, Phe-393, Pro-395, Pro-396, and Phe-397.

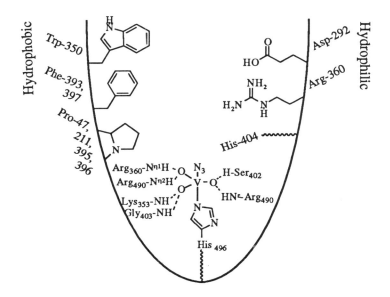

Fig. 2

Vanadate is bound with pentagonal bipyramidal geometry, ligated by azide (a result of the azide-containing crystallization buffer), three non-protein oxygen atoms, and histidine-496 (Fig. 3). The azide ligand is reportedly replaced by hydroxide in the native structure [26]. Vanadium coordination to the protein is stabilized by multiple hydrogen bonding between the vanadate oxygen atoms and the positively charged protein residues Lys-353, Arg-360, Arg-490, and Ser-402, as well as the amide nitrogen proton of Gly-403. Messerschmidt and Wever [26] propose that the hydrophobic residues Trp-350 and Phe-397 form a chloride binding site along with His-404; a hydrophobic binding site for halides is observed in other proteins such as haloalkane dehalogenase [27] and certain amylases [28]. His 404, which is present in the active site channel, must be deprotonated for H_2O_2 to bind to V-ClPO [21], and thus it is thought to function in acid-base catalysis.

Structural features of V-BrPO have not been published, although crystallization of V-BrPO from *A. nodosum* and *Corallina officinalis* has been reported at 2.4 Å and better than 2 Å resolution, respectively [12, 29]. The full sequence of V-BrPO (*A. nodosum*) has also not been reported, but from the partial sequence results [23] it is clear that there is sequence similarity between V-ClPO (*C. inaequalis*) and V-BrPO (*A. nodosum*), particularly in the active site region. Thus one can get an idea of the structure of V-BrPO. The similarities include regions containing four of the five amino acid residues which hydrogen bond to the vanadate oxygens (i.e. Arg-360, Ser-402, Gly-403, and Arg-490), the histidine ligand (His-496), and the acid-base histidine (His-404) [18, 23, 26]. In the proposed halide binding site [26], Trp-350 is present in both V-BrPO and V-ClPO, but Phe-397 is replaced by a histidine residue in V-BrPO, raising questions about the basis of halide specificity of each enzyme.

Extended X-ray absorption fine structure (exafs) analysis [30] and bond valence sum analysis [31, 32] are consistent with a trigonal bipyramidal vanadium site in V-BrPO (*A. nodosum*). These results along with the sequence similarities between V-BrPO and V-ClPO suggest that the vanadium site in V-BrPO has a trigonal bipyramidal structure, like that observed in V-ClPO, with coordination to one histidine ligand and multiple hydrogen bonds between the vanadate oxygens and positively charged residues.

The native vanadium(V) sites of V-BrPO and V-ClPO can be reduced to vanadium(IV) derivatives [19,30] which have nearly identical EPR signals, showing axially symmetrical vanadyl sites [19, 22]. While sequence similarities

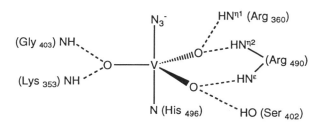

Fig. 3

between V-BrPO and V-ClPO suggest that the active sites of the two enzymes are similar, exafs analysis [30] and electron spin echo results [33] of reduced V-BrPO are consistent with a distorted octahedral geometry about the vanadium(IV) and an equatorial rather than an axial nitrogen ligand. These results imply either that the coordination environment of the vanadyl derivative of V-BrPO differs from the native structure, or that the coordination geometry of V-BrPO is different from that of V-ClPO.

3
Reactivity of the Vanadium Haloperoxidases

3.1
Halogenation and Halide-Assisted Disproportionation of Hydrogen Peroxide

Vanadium haloperoxidases (V-HPO) catalyze the halide-assisted disproportionation of hydrogen peroxide and the peroxidative halogenation of organic substrates. The enzymes catalyze the oxidation of the halide by hydrogen peroxide producing a two-electron oxidized halogen intermediate, such as hypobromous acid, bromine, tribromide, or an enzyme-bound bromonium ion equivalent, for bromide. This oxidation is followed by the halogenation of an appropriate organic substrate by the oxidized halogen intermediate, or a reaction with a second equivalent of hydrogen peroxide to form dioxygen, as shown in Scheme 1 for V-BrPO (*A. nodosum*) with bromide [34, 35].

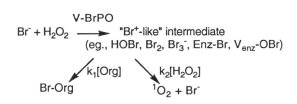

Scheme 1

Several features of this scheme deserve further comment:

Firstly, when monochlorodimedone (MCD) is the organic substrate, the reactive intermediate in Scheme 1 (e.g. HOBr, Br_2, Br_3^-, Enz-Br) is shown to be common to both the substrate halogenation and the dioxygen formation pathways [34,35]. The supporting evidence is provided by the nearly identical rates of monochlorodimedone bromination (>75 mM MCD) and dioxygen formation in the absence of organic substrate [34], as well as the kinetic parameters resulting from the steady-state enzyme kinetics [35]. The kinetic parameters (K_m^{Br}, $K_m H_2O_2$, K_{ii}^{Br}, K_{is}^{Br}; see below for further discussion of the enzyme kinetics) obtained in the MCD bromination reaction and the dioxygen formation reaction agree to within a factor of two, providing evidence that the rate-limiting steps are the same for both the bromination of MCD and the bromide-assisted disproportionation of hydrogen peroxide. In addition, $k_2[H_2O_2]$ competes with $k_1[MCD]$, because at variable concentrations of MCD or H_2O_2, the sum of $k_1[MCD]$ and $k_2[H_2O_2]$ in the presence of MCD is equal to $k_2[H_2O_2]$ in the absence of an organic halogen acceptor [36]. Competitive dioxygen formation is strongly enhanced at higher pH.

Secondly, the dioxygen formed through the V-BrPO-catalyzed disproportionation of hydrogen peroxide assisted by bromide oxidation is in the singlet excited state ($^1O_2$; $^1\Delta_g$), which was identified spectroscopically using the near-IR emission characteristics [13]. It has been well established that singlet oxygen ($^1\Delta_g$) is a product of the oxidation of H_2O_2 by certain oxidized halogen species (e.g. HOBr, HOCl, and bromamines) [37, 38]. The finding that each atom of oxygen in dioxygen originates from the same molecule of hydrogen peroxide, as shown in recent $H_2{}^{18}O_2$ studies [36], is also consistent with singlet oxygen production.

Thirdly, the V-BrPO-catalyzed bromination reaction has been shown to be an electrophilic (Br^+) rather than a radical ($Br\cdot$) process [36]. Product analysis of the bromination of 2,3-dimethoxytoluene (DMT) at pH 6.5 and 4 showed only a single product, ring-brominated 2,3-dimethoxytoluene. This product is indicative of electrophilic bromination. The product expected from a radical bromination process, 2,3-dimethoxybenzyl bromide, was not observed. In addition, two-electron oxidation of bromide is consistent with the formation of singlet oxygen.

3.2
Nature of the Oxidized Halogen Intermediate and the Question of Organic Substrate Binding to V-BrPO.

The nature of the oxidized intermediate, enzyme-bound or released, and the role of organic substrate binding to V-BrPO is not addressed in Scheme 1, nor in the experimental results described above. The oxidized halogen species cannot be detected because its reaction with organic substrates or with excess H_2O_2 to produce O_2 is too fast, preventing its build-up in solution. In competitive kinetic studies comparing the reactivity of the V-BrPO/H_2O_2/KBr system with HOBr, we have recently demonstrated that the nature of the halogenating species produced by V-BrPO depends on the nature of the organic substrate [38]. V-BrPO does not release an oxidized bromine species (e.g. HOBr, Br_2, Br_3^-) in the presence of certain indole derivatives, because these indoles bind to V-BrPO. (Halogenated indoles are marine natural products: see Fig. 1.) Indole binding was evaluated, in part, by fluorescence quenching experiments [38]. The Stern-Volmer analysis revealed a binding constant for 2-phenylindole to V-BrPO (*A. nodosum*) of 1.1×10^5 M^{-1} [39]. A mechanistic scheme involving substrate binding is shown in Scheme 2 [38]; V-BrPO binds H_2O_2 and Br^- leading to

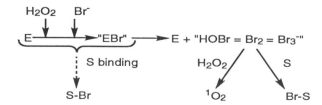

Scheme 2

a putative enzyme-bound or active-site trapped brominating moiety, E-Br, which in the absence of an indole may release HOBr (or other bromine species, e.g. Br_2, Br_3^-). When indole is present, it binds to V-BrPO, preventing release of an oxidized bromine species and leading to indole bromination.

The overall reactivity of V-ClPO (*C. inaequalis*) has not been investigated as extensively as has V-BrPO (*A. nodosum*). However, one important difference is apparent. The nature of the oxidized halogen intermediate for V-ClPO (*C. inaequalis*) can be addressed because it accumulates in solution under turnover conditions (e.g. 0.2 mM H_2O_2, 1 mM Cl^-, 64 nM V-ClPO in 0.1 M phosphate buffer, pH 4.5) [21]. V-ClPO generates HOCl, which can be detected spectrophotometrically. In addition, HOCl could be separated from the reaction solution by ultrafiltration (30,000 MW membrane). Both of the experiments were carried out at pH 4.5, which means that the reduction of HOCl with excess H_2O_2 producing dioxygen is not efficient at this pH. Since only about 50 mM HOCl was detected starting with an initial concentration of 200 mM H_2O_2, it was inferred that some reduction of HOCl by H_2O_2 could have occurred. Thus the overall reactivity in Scheme 1 probably holds for V-ClPO with chloride.

V-BrPO (*A. nodosum*) also catalyzes the peroxidation of chloride, although a much larger concentration of chloride is required (> 1 M [40]) compared to V-ClPO. As with the reactivity of V-BrPO (*A. nodosum*) with bromide, the rate of peroxidative chlorination of MCD is equivalent to the rate of the chloride-assisted disproportionation of hydrogen peroxide forming dioxygen (i.e., Scheme 1). However, in the presence of primary, secondary or tertiary amines, the rate of dioxygen formation is much slower due to the formation of chloramines, NH_3, NH_2R, NHR_2, NR_3 (R is alkyl, etc.) which do not readily oxidize hydrogen peroxide [11].[1] The formation of chloramines can be observed spectrophotometrically using their characteristic UV absorption maxima.

The rate of chlorination of MCD catalyzed by V-BrPO (*A. nodosum*) is the same in the presence and absence of an amine [40]. However, because the rate of chlorination of the organic substrate by the chloramine is very rapid, we were unable to determine whether MCD chlorination proceeds through a chloramine intermediate when an amine is present. Chloride oxidation seems to be most favorable at about pH 5 under conditions of 1 mM H_2O_2, 1.5 M KCl, 50 M MCD in 0.1 M citrate buffer, pH 5.0 [40].

3.3
The Steady-State Kinetics of V-BrPO and V-ClPO

Steady-state kinetic analyses of the rates of MCD bromination and dioxygen formation catalyzed by V-BrPO from *A. nodosum* [35, 41], *M. pyrifera* [11], and *F. distichus* [11] fit a substrate-inhibited bi bi ping-pong mechanism. At pH 6.5, the K_m^{Br} and $K_m^{H_2O_2}$ are ca. 22 mM and ca. 100 mM, respectively, for V-BrPO (*A.*

---

[1] Bromamine formation is not observed with V-BrPO using hydrogen peroxide [11] since bromamines are rapidly reduced by H_2O_2 to singlet oxygen and bromide [37]. Using acyl peracids as the oxidant instead of hydrogen peroxide, bromamines were observed [11]; see the section on peroxide reactivity.

nodosum).[2] Although bromide and hydrogen peroxide are substrates for V-BrPO, they are also both non-competitive inhibitors of V-BrPO in certain pH ranges. Bromide inhibition is strongest at pH 5–5.5 for V-BrPO from the three sources examined; K_i^{Br-} is 200–400 mM [11, 35, 41]. Hydrogen peroxide inhibition increases with increasing pH, ranging from $K_{ii}^{H_2O_2}$ of 312 mM at pH 5.5 to 64 mM at pH 8 [36].[3] Fluoride is a competitive inhibitor of H_2O_2 binding in both the MCD bromination reaction and the bromide-assisted disproportionation of hydrogen peroxide in V-BrPO: K_i^F is 1.75 mM at pH 6.5 [35]. Fluoride inhibition (K_i^F) is uncompetitive with respect to bromide [35].

The MCD chlorination kinetics for V-ClPO (*C. inaequalis*) also fit a substrate-inhibited bi bi ping-pong mechanism. The kinetic constant for chloride, K_m^{Cl}, is reported to be 0.25 mM at pH 4.5 [20]. The kinetic constant for hydrogen peroxide, $K_m^{H_2O_2}$, varies as a function of pH: 0.5 mM at pH 3.2 to 0.01 mM at pH 5 [21]. As with V-BrPO (*A. nodosum*), chloride is both a substrate for and an inhibitor of V-ClPO [20, 21]. At pH 3.1, chloride inhibition is competitive: K_i^{Cl} is 6 mM. At pH 4.1 chloride inhibition becomes non-competitive. Nitrate also inhibits V-ClPO competitively with respect to chloride and uncompetitively with respect to hydrogen peroxide at pH 5.5. The $K_i^{nitrate}$ is 2 mM at pH 5.5.

The kinetic studies of both V-BrPO and V-ClPO enzymes suggest that a residue with pK_a of 5.7–6.5 is important for reactivity [21, 35, 36, 41]. Under peroxide inhibition conditions of V-BrPO (*A. nodosum*), the steady-state kinetic constants indicate that an ionizable group with a pK_a between 6.5–7 is involved in the inhibition [36]. This residue must be deprotonated prior to binding of hydrogen peroxide. It seems reasonable to assume, given these pK_a values, that this residue is histidine, i.e. specifically the histidine in V-BrPO analogous to His 404 in V-ClPO.

3.4
Reversible and Irreversible Inactivation of V-BrPO under Turnover Conditions

H_2O_2 is a fully reversible, non-competitive inhibitor of V-BrPO. This is shown by the steady-state kinetics observed during the initial portion of the reaction (e.g. first ten minutes at the highest concentrations of H_2O_2) [36]. If the reaction proceeds longer, the specific activity of V-BrPO decreases. The inactivation that occurs on prolonged turnover between pH 6 and 8 can be fully reversed by the addition of vanadate, suggesting that protein-bound vanadium is released under turnover at high concentrations of hydrogen peroxide, forming the apo-enzyme derivative. Vanadium release was confirmed by atomic absorption analysis. At pH 4 and 5, inactivation also occurs under turnover, accompanied by the release of vanadium, but this turnover is not reversed on addition of vanadate.

---

[2] The K_m values reported are the average of the K_m values for the MCD bromination reaction and the dioxygen formation reaction.

[3] Inhibition by hydrogen peroxide could be investigated only using the bromide-assisted dioxygen formation reaction because the kinetics of the MCD bromination reaction were complicated by competing dioxygen formation at the high concentrations of hydrogen peroxide required for inhibition [36].

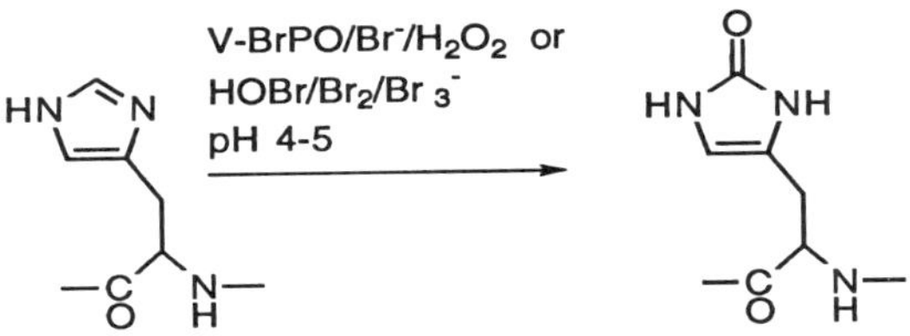

Fig. 4

The irreversible inactivation of V-BrPO which occurs at low pH was found to produce 2-oxohistidine as identified by HPLC using electrochemical detection [42]. Inactivation of V-BrPO and formation of 2-oxohistidine require all the components of turnover (i.e. bromide, hydrogen peroxide and V-BrPO) as well as low pH (Fig. 4). Neither inactivation nor 2-oxohistidine formation is observed in the presence of hydrogen peroxide alone. The inactivation and 2-oxohistidine formation are not the result of oxidation by singlet oxygen produced by V-BrPO, since they do not occur under conditions in which V-BrPO produces singlet oxygen quantitatively [42].

The oxidation of histidine to 2-oxohistidine is mediated by aqueous bromine species, as confirmed by the formation of N^{α}-benzoyl-2-oxohistidine upon addition of aqueous bromine to N^{α}-benzoylhistidine at low pH (Fig. 4). When hypobromite is added to N^{α}-benzoylhistidine in the presence of hydrogen peroxide at neutral pH, conditions under which HOBr would react with H_2O_2 to form $^1O_2$, the formation of N^{α}-benzoyl-2-oxohistidine is not observed. In addition, several functional mimics of V-BrPO (cis-VO_2^+ in strong acid, $MoO(O_2)_2(oxalato)_2$- at pH 5; see below) catalyze the bromide and hydrogen peroxide-mediated oxidation of N^{α}-benzoylhistidine to N^{α}-benzoyl-2-oxohistidine [42].

3.5
Reactivity of V-BrPO with Pseudohalides

In addition to iodide, bromide and chloride, V-BrPO catalyzes the oxidation of pseudohalides such as thiocyanate, azide, and cyanide [43, 44]. The reduction potentials for the oxidation of the pseudohalides fall within or below the range of those for the oxidation of I-, Br-, or Cl- (i.e, ε^o 0.53 V, 1.08 V, 1.36 V (vs. NHE) respectively). This suggests that vanadium haloperoxidases may also catalyze the oxidation of these anions. Indeed, in the presence of cyanide (ε^o = 0.375 V) or thiocyanate (ε^o = 0.77 V) the V-BrPO-catalyzed oxidation of bromide by hydrogen peroxide is inhibited [43, 45]. Once these pseudohalides are consumed, the V-BrPO-catalyzed peroxidation of bromide occurs at its normal rate. V-BrPO catalyzes the thiocyanation of 1,2-dimethylindole and 1,3,5-trimethoxybenzene to 1,2-dimethyl-3-thiocyanato-indole and 1-thiocyanato-2,4,6-trimethoxybenzene, respectively. ^{13}C NMR of the oxidation of $KS^{13}CN$ (133.4 ppm vs. TMS) by H_2O_2 catalyzed by V-BrPO showed the production of several oxidized thiocyanate species, including hypothiocyanate (OSCN-; 128.6 ppm), thio-oxime (SCNO- or SCNOH; 156.5 ppm), and the putative dithiocyanate ether anion (NCS-O-SCN; 127.6 ppm) which is unstable, as well as bicarbonate (HCO_3^-; 160.2 ppm). Azide, however, is a mechanism-based inactivator of V-BrPO [44, 46].

3.6
Peroxide Reactivity of V-BrPO

In place of hydrogen peroxide as the oxidant, V-BrPO can use acyl peracids (i.e. peracetic acid, phenyl peracetic acid, p-nitroperoxybenzoic acid, m-chloro-peroxybenzoic acid), but not alkylhydroperoxides (i.e. $tert$-butyl hydroperoxide, ethyl hydroperoxide, cuminyl hydroperoxide) [11]. The peracid reactivity results from direct use of the peracid and not from hydrogen peroxide which is a result of peracid hydrolysis. Bromamines (BrNHR) have been proposed as possible intermediates in other haloperoxidase reactions, and the formation of bromamines is observed when the reaction of V-BrPO with acyl peroxides is carried out in the absence of organic substrates and in the presence of amines (including buffers or amino acids). Since acyl peroxides are not efficient two-electron reductants, BrNHR species can accumulate in solution. The peracids do not readily reduce the oxidized bromine species (e.g. HOBr, Br_2, Br_3^-, BrNHR) in the time-frame of the enzymatic reactions (i.e. several minutes), thus dioxygen is not formed [11].

4
Functional Mimics of the Vanadium Haloperoxidases

4.1
cis-Dioxovanadium(V) in Acidic Aqueous Solution and Other Vanadium(V) Species.

The first reported functional mimic of vanadium bromoperoxidase was cis-dioxovanadium(V) (VO_2^+) in acidic aqueous solution [47, 48]. cis-Dioxovanadium(V) functions by coordination of hydrogen peroxide forming the corresponding oxomonoperoxo- and oxodiperoxo-vanadium(V) species (i.e. $VO(O_2)^+$ and $VO(O_2)_2^-$, respectively). The relative amounts of $VO(O_2)^+$ and $VO(O_2)_2^-$ produced depend on the acidity of the medium and the concentration of hydrogen peroxide according to the following equilibria:

$$VO_2^{2+} + H_2O_2 = VO(O_2)^+ + H_2O \qquad\qquad K_1 \quad (3)$$

$$VO(O_2)^+ + H_2O_2 = VO(O_2)_2^- + 2H^+ \qquad\qquad K_2 \quad (4)$$

where K_1 and K_2 are $(3.7 \pm 0.4) \times 10^4$ M^{-1} and 0.6 ± 0.1 M at 25 °C, respectively [47]. In general, as the pH is raised, the relative amount of $VO(O_2)^+$ decreases. In fact at neutral pH, only $VO(O_2)_2^-$ is present when at least two equivalents of hydrogen peroxide are added to a solution of vanadate. In acidic medium (e.g. 0.05 M H^+) containing ca. 5 mM H_2O_2 and 0.5 mM vanadium(V), both $VO(O_2)^+$ and $VO(O_2)_2^-$ are present [48]. Neither of these species oxidize bromide [49] (see below for oxidation of iodide), but rather they associate forming a binuclear oxovanadium(V), triperoxo- species, $(VO)_2(O_2)_3$, as supported by a new ^{51}V NMR signal, at 670 ppm (vs $VOCl_3$) [49, 50] (when carried out at a much higher vanadium concentration: 40 mM) and kinetic evidence [49]:

$$VO(O_2)^- + VO(O_2)_2^- = (VO)_2(O_2)_3 \qquad\qquad K_3 \qquad (5)$$

Despite the small equilibrium constant for formation of this triperoxo-, bis-oxovanadium(V) species (i.e. $K_3 = 9$ M^{-1} at pH 0–2), we have recently shown that it is the sole oxidant of bromide as depicted in Scheme 3 [49].

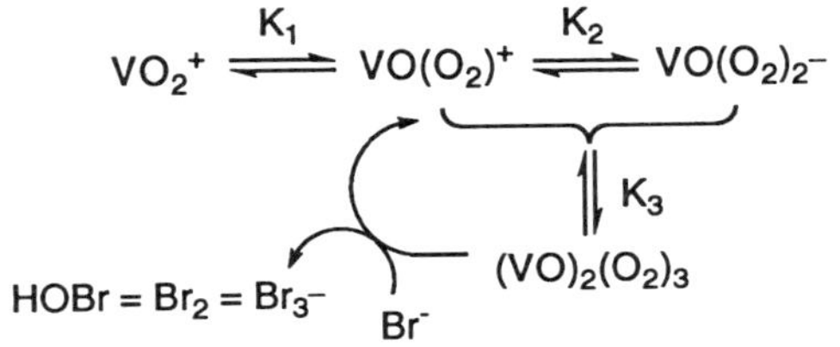

Scheme 3

The oxidation of bromide by $(VO)_2(O_2)_3$ produces HOBr, which is in a rapidly established equilibrium with bromine (Br_2) and tribromide (Br_3^-). In the absence of an organic substrate and in an acidic medium, the predominant species in solution is Br_3^-, which can be observed spectrophotometrically.[4] In the presence of an organic substrate, including MCD and trimethoxybenzene, the brominated product is formed quantitatively [48, 49].

The bimolecular rate constant for the oxidation of bromide by $(VO)_2(O_2)_3$ is 4.1 ± 0.1 M^{-1}s^{-1} [49]. Our kinetic analysis reveals that this bimolecular reaction is independent of acid, although it is required for the reaction to proceed: the acid participates in the overall catalytic cycle by maintaining the speciation of both vanadium(V) (i.e. cis-dioxovanadium(V) vs vanadate: $H_2VO_4^-$, HVO_4^{2-}, VO_4^{3-}) and oxoperoxo-vanadium(V) (i.e. the presence of both $VO(O_2)^+$ and $VO(O_2)_2^-$).

Wever estimated a lower limit for the bimolecular rate constant of the peroxo-V-BrPO complex with bromide [41]. This value, $k_{cat}/K_m^{H_2O_2}$, is strongly pH-dependent, primarily because $K_m^{H_2O_2}$ is sensitive to pH. The calculated bimolecular rate constant varies from 2.78×10^3 M^{-1}s^{-1} at pH 7.9 to 1.75×10^5 M^{-1}s^{-1} at pH 4.0, which is 10^3–10^5 times larger than the rate constant (4.1 ± 0.1 M^{-1}s^{-1}) for the reaction of $(VO)_2(O_2)_3$ and bromide [41, 49]. The speciation of the peroxo-vanadium(V) oxidant may control some of the reactivity difference. However, clearly the protein ligand mediates enzyme catalysis via a monomeric vanadium(V) center at neutral pH and a much faster rate (see below for further discussion).

In contrast to the mechanism of oxidation of bromide by the $VO_2^+/H_2O_2/Br^-$ system in acid, which is mediated by $(VO)_2(O_2)_3$ [48], Secco has found that iodide is oxidized by $VO(O_2)^+$, $VO(O_2)_2^-$ and VO_2^+ or their protonated forms [52, 53].

---

[4] The high concentrations of bromide and acid stabilize tribromide with respect to HOBr and Br$_2$ [51].

The diperoxo- species are the most reactive. The protonated form (i.e., $HVO(O_2)_2$) has a rate constant which is 200 times greater than the unprotonated species, which, Secco rationalizes, is a result of the more positive charge of the protonated form. Secco has used this result to argue for nucleophilic attack of iodide on the vanadium(V) center as opposed to a vanadium-bound peroxo-oxygen atom, although no direct experimental results address this point.

Other groups have also examined the vanadium(V)-catalyzed oxidation of bromide by hydrogen peroxide in aqueous or aqueous/organic mixtures, although without examining the detailed speciation of the vanadium peroxo-species [54–60]. These reports have focussed more on the nature of the substrate brominated and the product distribution under different conditions. DiFuria and co-workers have suggested that the oxidation of bromide by peroxo-V-BrPO may occur in a hydrophilic portion of the enzyme, followed by migration of the oxidized bromine intermediate to a more hydrophobic portion of the protein where bromination of organic substrates occurs and where decomposition by hydrogen peroxide is limited [54, 55]. In support of this suggestion, this group has used a two-phase 0.1 M $HClO_4$-CH_2Cl_2 (or 0.1 M $HClO_4$-$CHCl_3$) system to show that bromine can be transferred and accumulate in the organic phase [54]. If the reaction is carried out with an organic substrate, quantitative bromination is observed.[5]

The two-phase system [54, 55] has low selectivity; dibromo-derivatives and bromohydrins are observed along with the mono-brominated products. However, the ratio of dibromo-derivatives and bromohydrins to mono-brominated product is different in the two-phase system with the vanadium catalyst than with bromination with Br_2 or HOBr in a two phase system [55]. These authors conclude that because the vanadium-oxidized bromine-equivalent intermediate behaves differently than bromine, a vanadium(V)-hypobromite-like species is formed and is an active brominating moiety. However, it is not clear whether the acid and bromide concentration of the two-phase system is the same as in the vanadium catalytic system [55]. Clearly the results of further investigations will be interesting.

An oxoperoxo-vanadium(V) complex of glycine $[V_2O_2(O_2)_3(GlyH)_2(H_2O)_2]$ which oxidizes bromide in water is interesting because it may be related to the $(VO)_2(O_2)_3$ complex described above. It was proposed that one of the peroxide ligands is present as a μ-peroxo-moiety, but X-ray structural analysis has not been reported [58].

---

[5] It should be noted in reference [55] that the aqueous phase is acidic which stabilizes Br_3^- over HOBr. Thus oxidation of excess H_2O_2 by the oxidized bromine species is not favored [54]. The point is made that quantitative bromination indicates that the side-reaction of oxidation of excess H_2O_2 by the bromine species is minimized, because bromine transfers easily to the organic phase but the hydrogen peroxide or the peroxo vanadium species remain in the aqueous phase. However, even in a purely aqueous medium, quantitative bromination of trimethoxybenzene was observed [48, 49]. Thus a suitable organic substrate competes favorably against hydrogen peroxide for the oxidized bromine species in acidic solution.

4.2
Molybdenum(VI), Tungsten(VI) and Rhenium(VII)

That peroxo-vanadium(V) species oxidize halides raises questions about the re-activity of other peroxo-metal complexes. Indeed, peroxidative halogenation has been shown to occur with molybdenum(VI) [61–64], tungsten(VI) [61], and methyl-rheniumtrioxide [65, 66]. However, aqueous titanium(IV) has been found to actually stabilize peroxide against oxidation of even iodide [67].

Molybdenum and Tungsten. The catalytic cycle for the aqueous species of molybdenum(VI) and tungsten(VI) is shown in Scheme 4 [61]. Oxodiperoxo-Mo(VI) and W(VI) are the only detectable species present in acidic solution (3 mM – 0.2 M $HClO_4$) in the presence of two or more equivalents of H_2O_2 per equivalent of Mo(VI) or W(IV). The rate of bromination of TMB is proportional to $[Br^-]$ and $[Mo]$ (or $[W]$) and zero-order in $[H_2O_2]$. The independence of the rate on $[H_2O_2]$ is consistent with rapid recoordination of H_2O_2 to Mo(VI) and a rate-limiting step of bromide oxidation. The fate of the monoperoxide species is not known: it could coordinate H_2O_2, oxidize another equivalent of bromide, or disproportionate, forming MoO_3 and $MoO(O_2)_2$. The pH dependence shows an inflection at pH 2.1 which is the pK_a of $(H_2O)_2MoO(O_2)_2$.[6]

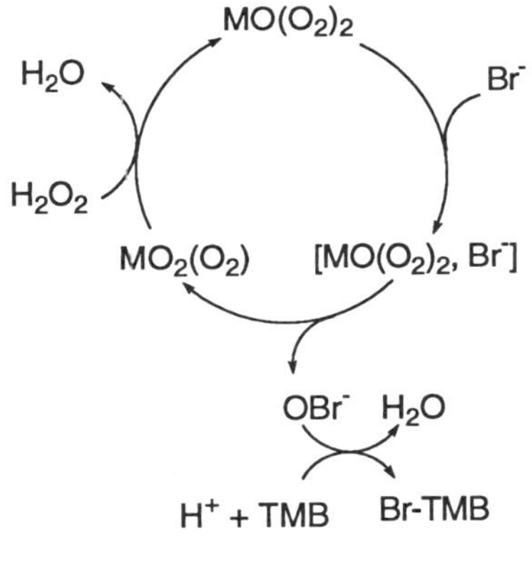

Scheme 4

The general reactivity for bromide oxidation is $WO(O_2)_2 > MoO(O_2)_2 > VO(O_2)_2^-$.[7] It has been suggested that the increase in activity of the group VI metals compared to vanadium(V) mimics may be due to their higher oxidation state, which would increase the oxidation potential of the bound peroxide [61]. In addition, a correlation has been drawn between the acidity of a complex and

---

[6] The inflection point in the acid dependence reaction with W(VI) could not be determined due to an increase in the rate of the uncatalyzed oxidation of bromide by hydrogen peroxide at H^+ concentrations higher than 0.1 M. The W(VI)-catalyzed reaction is faster at lower pH, consistent with the lower pK_a of 0.12 [61].

[7] In fact one must remember, as described above, that $VO(O_2)_2^-$ does not oxidize bromide directly [49].

its reactivity: under similar conditions, $V(V)$, $Mo(VI)$, and $W(VI)$ show relative reactivities of $10:10^4:10^5$ [68]. As mentioned above, $Ti(IV)$, which has less Lewis acid character, stabilizes peroxide [67]. In the case of $V(V)$, the slow rate of bromide oxidation when compared to $MoO(O_2)_2(H_2O)_2$ and $WO(O_2)_2(H_2O)_2$ under identical conditions (see Fig. 7 in reference [61]) is due to $(VO)_2(O_2)_3$ and not $VO(O_2)_2^-$. In fact $VO(O_2)_2^-$ does not oxidize bromide. $(VO)_2(O_2)_3$ is present in very low concentrations which is a result of its small formation constant (i.e., $9\ M^{-1}$) [49]. Nevertheless, the second-order rate constant for bromide oxidation by $(VO)_2(O_2)_3$ is faster than for $MoO(O_2)_2(H_2O)_2$ or $WO(O_2)_2(H_2O)_2$ (see Table 1 below).

The oxalato complex of oxodiperoxo-molybdenum(VI), $MoO(O_2)_2(C_2O_4)^{2-}$, also catalyzes the oxidation of bromide [61, 62]. The proposed mechanism involves cycling between the diperoxo-$MoO(O_2)_2(C_2O_4)^{2-}$ species and a mono-peroxo-intermediate, $MoO_2(O_2)(C_2O_4)^{2-}$, which combines with hydrogen peroxide to regenerate the initial diperoxo-complex [62]. In the case of $MoO_2(O_2)(C_2O_4)^{2-}$, which was studied at pH 5, dioxygen formation was observed in the absence of an organic substrate acceptor, whereas in the case of $MoO(O_2)_2(H_2O)_2$ in strong acid, dioxygen formation was not observed. This apparent difference in reactivity is a result of the speciation of HOBr, Br_2 and Br_3^- as a function of acid; thus Br_3^- which is favored in acid, does not oxidize H_2O_2, while at pH 5 more HOBr is present which does oxidize H_2O_2, forming dioxygen [69].

Chloride was found to inhibit the oxidation of bromide by the molybdenum(VI)-oxodiperoxo-complex [61]. The nature of the inhibition is not

Table 1. Bimolecular rate constants for the oxidation of bromide by peroxo-metal species

Oxidant	Conditions	Rate Constant $M^{-1}s^{-1}$	Reference
V-BrPO-(O_2)	pH 7.9[a]	2.78×10^3	41
	pH 4.0[a]	1.75×10^5	41
$MeReO_2(O_2)$	pH 0	350	65
$MeReO(O_2)_2$	pH 0	190	65
$VO(O_2)(Hheida)$	CH_3CN[b]	280	73
$VO(O_2)(bpg)$	CH_3CN[b]	21	73
$(VO)_2(O_2)_3$	pH 0.7–2.0	4.1	49
$MoO(O_2)_2(H_2O)_2$	pH 1.0–5.1	1.5×10^{-2}	61
$MoO(O_2)_2(H_2O)(OH)^-$	pH 1.0–5.1	2.4×10^{-3}	61
$MoO(O_2)_2(C_2O_4)^{2-}$	pH 5.1 with 20% MeOH	9.2×10^{-3}	61
	pH 5.0	4.9×10^{-3}	62
$VO(O_2)^+$	pH 0.7–2.0	nd	49
$VO(O_2)_2^-$	pH 0.7–2.0	nd	49

nd = not detected.
[a] lower limit as calculated by $k_{cat}/K_m^{H_2O_2}$.
[b] bromide oxidation did not occur in water.

clear but may be due to a change in the oxidation potential and the charge of the complex when it is bound to chloride. Chloride inhibition with $MoO(O_2)_2(H_2O)_2$ is accompanied by a shift in λ_{max} from 324 nM to 329 nM upon chloride addition, suggesting that chloride may coordinate to Mo(VI). A similar shift was not observed when only bromide was present, but this may be because the rate of bromide oxidation is too fast for a bromo-oxoperoxo-molybdenum(VI) intermediate to be observed. When chloride is added to the oxalato complex of $Mo(O_2)_2$, i.e. $MoO(O_2)_2(C_2O_4)^{2-}$, which does not have a dissociable ligand, no shift in the UV-vis spectrum of the complex was observed and there was no decrease in the rate of bromide oxidation [61]. Thus, halide coordination to the metal can occur, but it is not necessary for bromide oxidation.

Rhenium. Methylrhenium trioxide, CH_3ReO_3, has also been found to catalytically oxidize halides in acidic solution (Scheme 5) [65, 66]. The monoperoxo-complex, $CH_3ReO_2(\eta^2\text{-}O_2)$, and the diperoxo- complex, $CH_3ReO_2(\eta^2\text{-}O_2)_2(H_2O)$ both oxidize bromide and chloride in 1.0 M $HClO_4$. The reaction of the catalytic species (i.e, $CH_3ReO_2(\eta^2\text{-}O_2)$ + $CH_3ReO_2(\eta^2\text{-}O_2)_2(H_2O)$) with chloride is 10^5 times faster than the uncatalyzed reaction [66]. The rate of oxidation of bromide is at least 10^3 times faster than for chloride [65]. Espenson and Hansen [66] argue that the activity of monoperoxo-complexes in other systems should not be disregarded since the monoperoxo- to diperoxo-equilibria for these complexes are in excess of 10^6 M^{-1} [78,79] and thus the monoperoxo-complexes are in such low concentrations that their reactivity cannot be observed.

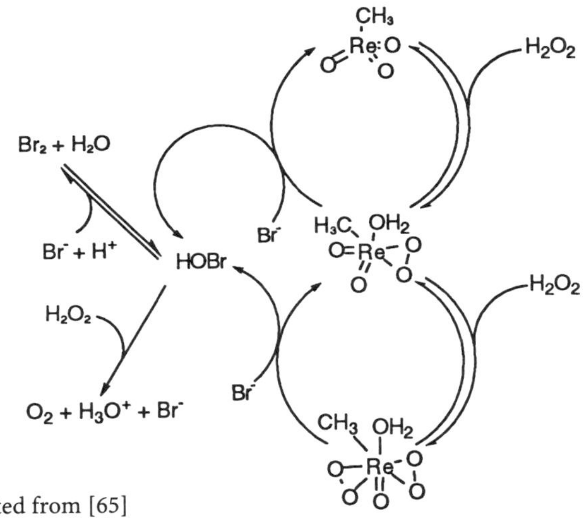

Scheme 5. Adapted from [65]

4.3
The Hydroxyphenyl-Salicylideneamine Complex of VO(OH)

Mononuclear vanadium(V) complexes also catalyze the oxidation of halides by hydrogen peroxide. One example is the (hydroxo)oxovanadium(V) complex of hydroxyphenyl-salicylideneamine (H_2HPS) [70]. (HPS)VO(OH) is one of several species formed when the starting material, (HPS)VO(OEt)(EtOH) (Fig. 5), is dissolved in DMF (Scheme 6). In addition to (HPS)VO(OH), four other complexes are formed: (HPS)VO(OEt), and three stereochemically distinct dimers, $[(HPS)VO]_2O$, as identified by ^{51}V NMR (Scheme 6) [71,72]. A single oxoperoxovanadium(V) species, $(HPS)VO(O_2)^-$, is formed on addition of H_2O_2 to the mixture. When bromide is added to this solution, one turnover stoichiometric with respect to the concentration of the vanadium complex is observed in the absence of added acid, stopping at $(HPS)VO_2^-$. If acid is added, so that it is at least stoichiometric with H_2O_2, bromination becomes catalytic in the vanadium complex. Bromination is also stoichiometric with respect to H_2O_2 consumption. Like V-BrPO and the functional mimics discussed above (cis-VO_2^+, Mo(VI), W(VI)), bromination of organic substrates was shown to occur by an electrophilic process (Br$^+$) and not a radical process (Br•) [71]. Thus with the HPS^{2-} complex, unlike cis-VO_2^+, the monomeric monoperoxo-vanadium(V) complex oxidizes bromide.

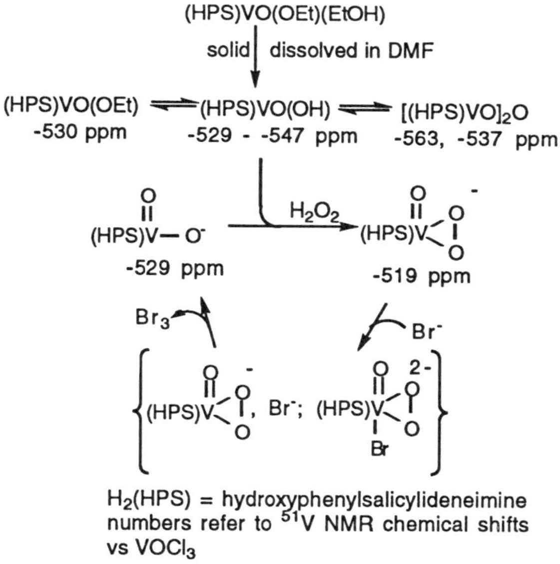

(HPS)VO(OEt)(EtOH)

solid | dissolved in DMF

(HPS)VO(OEt) ⇌ (HPS)VO(OH) ⇌ [(HPS)VO]$_2$O
-530 ppm   -529 - -547 ppm   -563, -537 ppm

(HPS)V—O$^-$ $\xrightarrow{H_2O_2}$ (HPS)V
-529 ppm   -519 ppm

Br$_3^-$ Br$^-$

{(HPS)V , Br$^-$; (HPS)V}

H_2(HPS) = hydroxyphenylsalicylideneimine
numbers refer to ^{51}V NMR chemical shifts
Scheme 6   vs VOCl$_3$

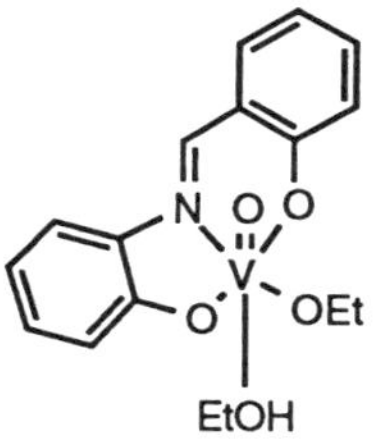

Fig. 5   VO(HPS)(OEt)(EtOH)

4.4
Other Complexes

Bromide oxidation by the peroxide complexes of several other V(V)-L complexes is similar to (HPS)VO(O$_2$) described above. Such complexes include the salicylidene-amino acid Schiff base ligands (H$_2$Sal:Phe, H$_2$Sal:Gly), iminodiacetic acid (H$_2$IDA), nitrilotriphosphoric acid (H$_3$NTP) and other tripodal amine chelates (see below) and citric acid (Fig. 6) [31, 73]. On the other hand, the vanadium(V) complex of pyridine-2,6-dicarboxylic acid (H$_2$dipic) stabilizes coordinated H$_2$O$_2$ against bromide oxidation. Finally, some ligands cannot compete with simultaneous coordination of H$_2$O$_2$; for example carboxyphenyl-salicylideneamine (H$_2$CPS) and pyridine-2-carboxylic acid (H$_2$pic) dissociate in the presence of H$_2$O$_2$ [31,71].

The oxoperoxo-vanadium(V) complexes of N-(2-hydroxyethyl)iminodiacetic acid (Hheida), N-(2-amidomethyl)iminodiacetic acid (ada), and N,N-bis(2-pyridylmethyl)-glycine (bpg) were shown to have a distorted pentagonal bipyramidal geometry [73, 74], consistent with the structures of all other oxo-peroxo-vanadium(V) complexes reported to date [72]. The reactivity of these complexes in acetonitrile is as described above for the other monomeric vanadium(V) complexes [70]. The bimolecular rate constants for bromide oxidation by the peroxo-complexes varied from 21 M^{-1}s^{-1} for the Hbpg complex to 280 M^{-1}s^{-1} for the Hheida complex (see below for further discussion of these rate constants) [73]. The rates for iodide oxidation were approximately 5–6 times faster than for bromide.

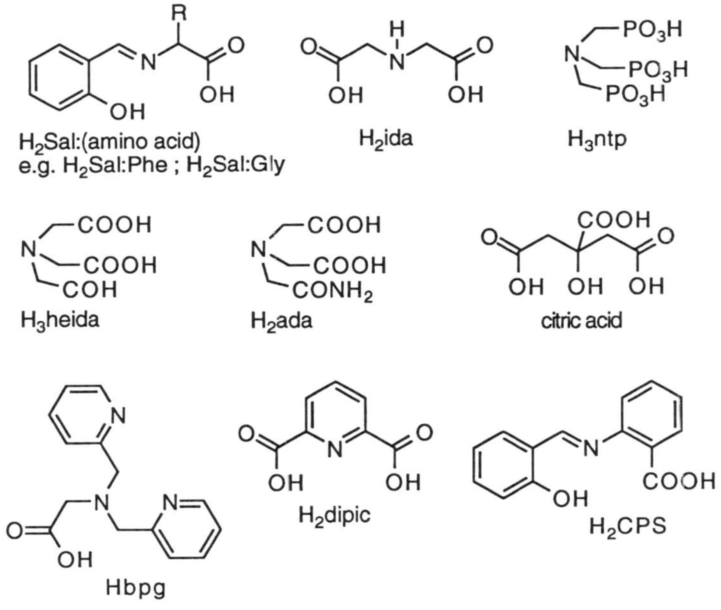

Fig. 6

5
On the Mechanism of V-HPO: Consideration of the Reaction Sequence, Role of Vanadium and Role of the Protein

Spectrophotometric analysis indicates that the first step in the catalytic cycle of V-HPO is coordination of hydrogen peroxide to vanadium(V) (Step 1 in Scheme 7). Upon addition of H_2O_2 to V-BrPO, a small absorbance decrease occurs between 300–340 nM [75]. Subsequent addition of bromide yields the original UV spectrum, a result consistent with bromide oxidation by an enzymatic peroxo-vanadium(V) species. In a separate experiment, it was found that oxidation of bromide by V(V)-BrPO, prior to reaction with hydrogen peroxide, could be ruled out: when excess bromide was added to V-BrPO (20 mM) and MCD (50 mM), MCD bromination was not observed, indicating that V-BrPO does not oxidize bromide in the absence of a peroxide source (hydrogen peroxide or an acyl peroxide) [36, 44].

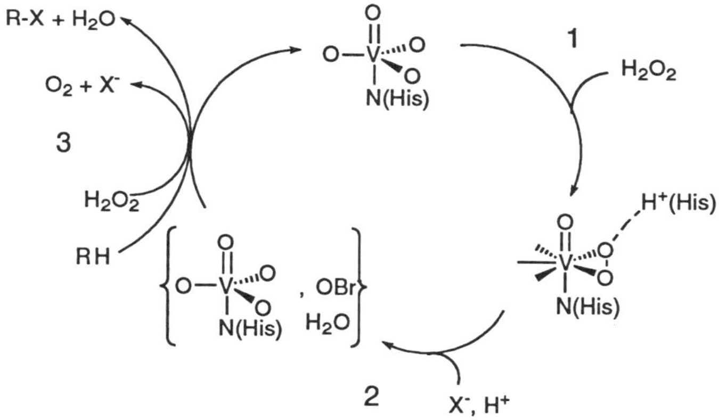

Scheme 7

The coordination mode of the peroxo-V-HPO complex is not known. From small molecule studies, it is known that all peroxo- vanadium(V) complexes, whether productive catalysts of halide oxidation reactions or not, contain η^2-coordinated peroxide (for a review see [72, 76]). Oxomonoperoxo-vanadium(V) complexes are characterized by symmetrical η^2-coordination of the peroxide ligand. Oxodiperoxo-vanadium(V) complexes contain a longer peroxo-O-O bond than the oxomonoperoxo-compounds and they are characterized by asymmetrical η^2-coordination of the peroxide: one M-O bond is longer in the diperoxo- complexes (i. e., the V-O$_{trans}$ bond) than the other (Fig. 7). Thus, one might expect the oxodiperoxo-complexes would be more reactive to oxidation. Indeed, $VO(O_2)_2^-$ oxidizes iodide faster than $VO(O_2)^+$ [47] and in general aqueous oxodiperoxo-vanadate species are more reactive in organic oxidation reactions than their monoperoxo-counterparts [51]. However, $VO(O_2)_2^-$ does not oxidize bromide. Thus the question is raised as to whether the reactivity of V-HPO arises from a different mode of coordination in the enzyme. For example is an η^1 or hydroperoxide species the active oxidant of the halide? Or does

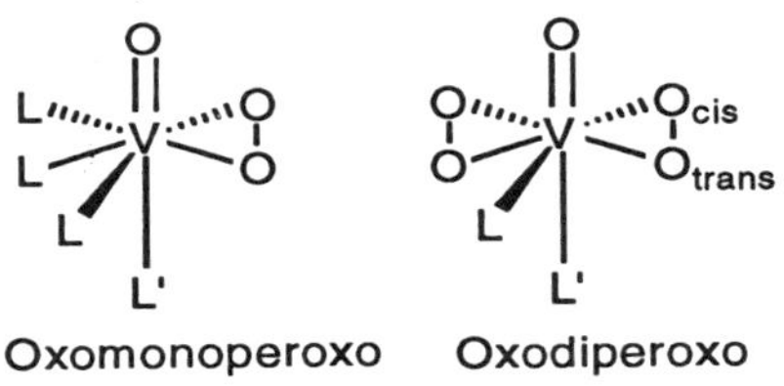

Oxomonoperoxo   Oxodiperoxo

Fig. 7

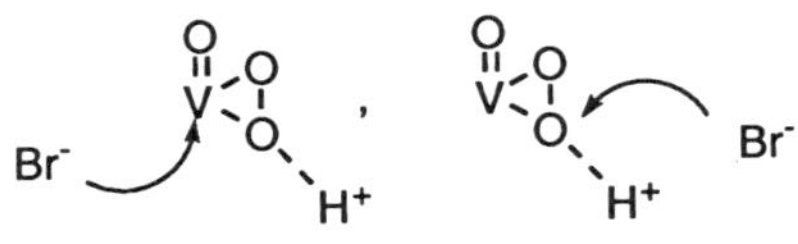

Fig. 8

the enzyme cycle between the diperoxo- and monoperoxo-species? The answers to these questions may be addressed in part by crystallographic analysis of the peroxo-V-HPO complexes.

After the first step of peroxide coordination to vanadium(V) in V-HPO, the halide is oxidized (Step 2 in Scheme 7). It has not been established conclusively whether the halide attacks the coordinated peroxide directly or whether it coordinates to vanadium(V) prior to oxidation (Fig. 8). From mechanistic studies with the molybdenum(VI) functional mimics, comparing $MO(O_2)_2(H_2O)$ with $MO(O_2)_2(C_2O_4)^{2-}$, it appears that a vacant coordination site on the oxoperoxo- metal species is not a prerequisite for halide oxidation. Thus, the oxidation probably occurs via nucleophilic attack by the halide on the coordinated peroxide.

Enzyme kinetics with V-BrPO and V-ClPO show saturation in bromide and chloride, establishing that a halide binding site exists, but the nature of this site remains unclear. Messerschmidt and Wever propose a chloride binding site of Trp-350, Phe-397 and His 404 in V-ClPO (*C. inaequalis*) based on similarity with other halide binding proteins [26]. All of these residues, however, are not conserved in V-BrPO (*A. nodosum*) (i.e., Phe is replaced by His). Thus, the basis of the preferred halide reactivity in V-HPO may be related to the nature of these amino acid residues. Of course the halide binding site could be vanadium(V), however, no difference in the exafs of V-BrPO (*A. nodosum*) was observed in the presence of bromide [80].

No direct evidence of an electron transfer role for vanadium has been detected in V-BrPO, which is surprising given the number of stable oxidation states available to vanadium in aqueous solution (see discussion in previous reviews: [14, 31, 45, 72, 76]). The reaction of V-BrPO is consistent with a Lewis acid role of vanadium(V), in which hydrogen peroxide coordination to vanadium(V) activates peroxide for halide oxidation. No epr signal has been observed under turnover conditions [41], and VO^{2+}-BrPO does not have bromoperoxidase activity, so the vanadyl-BrPO state is probably not an important component in the

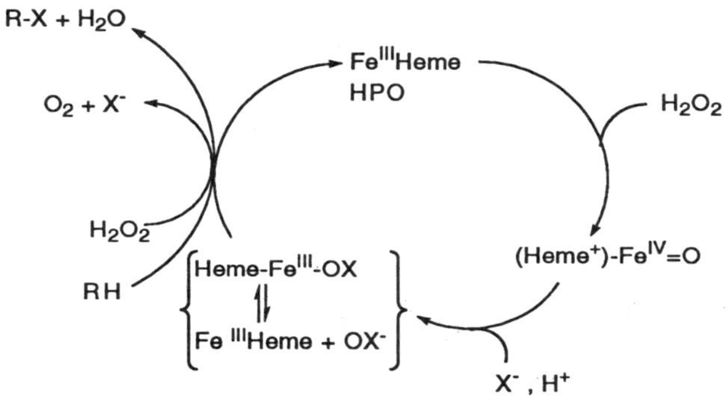

R-X + H$_2$O

O$_2$ + X⁻

Fe^{III}Heme
HPO

H$_2$O$_2$

H$_2$O$_2$

RH

$\begin{bmatrix} \text{Heme-Fe}^{III}\text{-OX} \\ \Updownarrow \\ \text{Fe }^{III}\text{Heme + OX}^- \end{bmatrix}$

(Heme$^+$)-Fe^{IV}=O

X⁻, H$^+$

Scheme 8

catalytic cycle. Also, a long-lived reduced oxidation state is not observed in ^{51}V NMR studies of functional model complexes [48, 49].

The FeHeme haloperoxidases function as two-electron redox catalysts (Scheme 8). Hydrogen peroxide oxidizes the Fe(III)-heme moiety by two electrons, forming Compound 1 (Heme$^+$-Fe^{IV} = O). Compound 1 then oxidizes the halide with two electrons, reforming the Fe(III)-heme moiety and a hypohalite species.

This sequence cannot be the mechanism in V-BrPO since vanadium is already in the highest oxidation state. Native V-BrPO (protein-V^V = O), which is analogous to Compound 1 (Heme$^+$-Fe^{IV} = O), does not oxidize bromide directly, as demonstrated by the lack of stoichiometric bromination when bromide was added to V-BrPO in the absence of hydrogen peroxide [36, 44]. An interesting inorganic system pertains to the discussion of oxidation state changes of the vanadium center in V-HPO as well as the models. The vanadium(V) complex of tetraethyleneglycol (H$_2$teg) oxidizes HBr in 1,2-dichloroethane forming a vanadium(III) complex, [V(teg)Br$_2$]Br, which was crystallographically characterized [77]. In the presence of air, V(III) is oxidized to vanadium(V). This reaction sequence differs from V-BrPO and the vanadium(V) catalysts discussed above where neither the reduction of vanadium(V) nor the oxidation of bromide is observed in the absence of hydrogen peroxide.

The second-order rate constants for the oxidation of bromide by various peroxo-metal species are summarized in Table 1. The peroxo-V-BrPO complex has the largest rate constant for the oxidation of bromide; thus the protein active site must provide a favorable environment. One feature of V-ClPO, which is probably present in V-BrPO, is the acid/base histidine (i.e., His 404 in V-ClPO from *C. inaequalis*). His 404 lies in hydrogen bonding distance of vanadium(V). Thus it probably assists in the reductive cleavage of coordinated peroxide upon nucleophilic attack by bromide. None of the model complexes has this feature. The relatively large rate constants for the oxidation of bromide by VO(O$_2$)(Hheida) and VO(O$_2$)(bpg) which were observed in acetonitrile, but not water, may be a result of the protonation of the η^2-coordinated peroxide which occurs in acetonitrile, but not water [73].

The state of the oxidized halogen intermediate, enzyme-bound or released, is affected by the nature of the organic substrate [38]. As discussed above, substrates which bind to V-BrPO (*A. nodosum*) block the release of a diffusible oxidized bromine intermediate. Thus one may anticipate that regioselective halogenation can be achieved with this enzyme. Indeed, this is an active area of research. Another active area, and one which when achieved will certainly advance our understanding of the mechanism of halogenation by the marine haloperoxidases, is that of stereoselective halogenation.

Vanadium has been used as a catalyst in many industrial processes. Here we have illustrated the catalytic role of a vanadium enzyme in a biological process. The rate of halogenation catalyzed by V-HPO at its physiological pH (pH 5–7) is faster than the functional mimics reported to date, despite the very simple vanadium site in the enzyme (i.e., vanadate coordination to the protein by one histidine ligand). The protein clearly plays a very important role in the activation of vanadium-bound peroxide towards halide oxidation. We eagerly await the results of future mechanistic studies as well as novel applications of this robust and active class of enzymes.

Acknowledgements: Alison Butler gratefully acknowledges support from the National Science Foundation (CHE96-29374; MCB90-18025. A.B. is an Alfred P. Sloan Research Fellow. Partial support for this work is also sponsored by NOAA, U.S. Department of Commerce under grant numbers NA89-D-SG138, project number R/MP-53 and NA66RGO477 project number R/MP-76 through the California Sea Grant College System, and in part by the California State Resources Aqency. The views expressed herein are those of the authors and do not necessarily reflect the views of NOAA. The U.S. Government is authorized to reproduce and distribute for governmental purposes.

6
References

1. Hager LP, Morris DR, Brown FS, Eberwein H (1966) J Biol Chem 241:1769
2. Gschwend PM, MacFarlane JK, Newman A (1985) Science, 227:1033
3. Walter B, Ballschmiter K (1991) Chemosphere 22:557
4. Manley SL, Goodwin K, North WJ (1992) Limnology And Oceanography 37:1652
5. Faulkner DJ (1993) Natural Product Reports, 10:497 and references therein
6. Vollenbroek EGM, Simons LH, van Schijndel JWPN, Barnett P, Balzer M, Dekker H, van der Linden C, Wever R (1995) Biochem Soc Trans 23:267
7. Krenn BE, Tromp MGM, Wever R (1989) J Biol Chem 264:19287
8. de Boer E. Tromp MGM, Plat H, Krenn BE, Wever R (1986) Biochim Biophys Acta 872:104
9. Wever R, Krenn BE, de Boer E, Offenberg H, Plat H (1988) Prog Clin Biol Res (Oxidases Relat Redox Syst) 274:477
10. de Boer E, van Kooyk Y, Tromp MGM, Plat H, Wever, R (1986) Biochim Biophys Acta 869:48
11. Soedjak HS, Butler A (1990) Biochemistry 29:7974
12. Muller-Fahrnow A, Hinrichs W, Saenger W, Vilter H (1988) Febs Lett 239:292
13. Everett RR, Kanofsky JR, Butler A (1990) J Biol Chem 265:4908
14. Butler A, Walker JV (1993) Chem Rev 93:1937
15. de Boer E, Plat H, Tromp MGM, Wever , Franssen MC, Van der Plas HC, Meijer EM, Schoemaker HE (1987) Biotechnol Bioeng 30:607
16. Sheffield, DJ, Smith AJ, Harry TR, Rogers LJ (1993) Biochem Soc Trans 21:445S
17. Itoh N, Cheng LY, Izumi Y, Yamada H (1987) J Biotech 5:29
18. Simons LH, Barnett P, Vollenbroek EGM, Dekker HL, Muijsers AO, Messerschmidt A, Wever R (1995) Eur J Biochem 229:566

19. van Schijndel JWP, Vollenbroek, EGM, Wever R (1993) Biochim Biophys Acta 1161:249
20. van Schijndel JWPM, Simons LH, Vollenbroek EGM, Wever R (1993) FEBS Lett 336:239
21. van Schijndel JWPM, Barnett P, Roelse J, Vollenbroek EGM, Wever R (1994) Eur J Biochem 225:151
22. de Boer E, Boon, K, Wever R (1987) Biochemistry 27:1629
23. Vilter, H (1995) Metal Ions in Biological Systems 31:325
24. Soedjak HS, Butler A (1991) Biochim Biophys Acta 1079:1
25. Tromp M, Tran TV, Wever R (1991) Biochim Biophys Acta 1079:53
26. Messerschmidt A, Wever R (1996) Proc Natl Acad Sci, USA, 93:392
27. Verschueren KHG, Kingma J, Rozeboon HJ, Kalk KH, Janssen DB, Dijkstra BW (1993) Biochemistry 32:9031
28. Machius M, Wiegand G, Huber R (1995) J Mol Biol 246:545
29. Rush C, Willetts A, Davies G, Dauter Z, Watson H, Littlechild J (1995) FEBS Letters, 359:244
30. Arber JM, de Boer E, Garner CD, Hasnain SS, Wever R (1989) Biochemistry 28:7968
31. Butler A, Clague MJ (1995) in Mechanistic Bioinorganic Chemistry, (ACS Symposium Volume, eds, HH Thorp and VL Pecoraro) 329
32. Carrano CJ, Mohan M, Holmes SM, de la Rosa RI, Butler A, Charnock JM, Garner CD (1994) Inorg Chem 33:646
33. de Boer E, Keijzers CP, Klassen AAK, Reijerse EJ, Collison D, Garner CD, Wever, R (1988) FEBS Lett 235:93
34. Everett RR, Butler A (1989) Inorg Chem 28, 393
35. Everett RR, Soedjak HS, Butler A (1990) J Biol Chem 265:15671
36. Soedjak HS, Walker JV, Butler A (1995) Biochemistry 34:12689
37. Kanofsky JR (1989) Arch Biochem Biophys 274:229
38. Tschirret-Guth RA, Butler A (1994) J Am Chem Soc 116:411
39. Tschirret-Guth RA (1996) PhD Dissertation, UC Santa Barbara, p 223
40. Soedjak HS, Butler A (1990) Inorg Chem 29:5015
41. de Boer E, Wever R (1988) J Biol Chem 263:12326
42. Winter GEM, Butler A. (1996) Biochemistry 35:11805
43. Walker JV, Butler A (1996) Inorg Chim Acta 243:201
44. Soedjak HS (1991) PhD Dissertation, UC Santa Barbara, p 283
45. Wever R, Kustin K (1990) Adv Inorg Chem 35:81
46. Everett RR (1990) PhD Dissertation. UC Santa Barbara, p 211
47. Secco F (1980) Inorg Chem 19:2722
48. de la Rosa RI, Clague MJ, Butler A (1992) J Am Chem Soc, 114:760
49. Clague MJ, Butler A (1995) J Am Chem Soc 117:3475
50. Harrison AT, Howarth OW (1985) J Chem Soc Dalton Trans 1985:1173
51. Thompson RC (1986) Adv Inorg Bioinorg Mech 4:65
52. Secco F, Celsi S, Grati C (1972) JCS Dalton 1972:1675
53. Celsi S, Secco F, Venturini M (1974) JCS Dalton 1974:793
54. Conte V, DiFuria F, Moro S (1994) Tet Lett, 35:7429
55. Andersson M, Conte V, DiFuria F, Moro S (1995) Tet Lett, 36:2675
56. Bhattacharjee M (1992) Polyhedron 11:2817
57. Bhattacharjee M, Ganguly S, Mukherjee J (1995) J Chem Res, 1995:80
58. Bhattacharjee M, Chaudhuri MK, Islam NS, Paul PC (1990) Inorg Chim Acta 169:97
59. Dinesh C, Kumar R, Pandey B, Kumar P (1995) Chem Commun 1995:611
60. Hegde VR, Pais GCG, Kumar R, Kumar P, Pandey, B (1996) J Chem Res-S, 1996:62
61. Meister GE, Butler A (1994) Inorg Chem 33:3269
62. Reynolds MS, Morandi SJ, Raebiger JW, Melican SP, Smith SPE (1994) Inorg Chem 33:4977
63. Smith RH, Kilford J (1976) International Journal of Chemical Kinetics 8:1
64. Arias C, Mata F, Perez-Benito JF (1990) Can J Chem 68:1499
65. Espenson JH, Pestovsky O, Hansen P, Staudt S (1994) J Am Chem Soc 116:2869
66. Hansen P, Espenson JH (1995) Inorg Chem 34:5839

67. Lydon J.D, Thompson RC (1986) Inorg Chem 25:3694.
68. Ghiron A F, Thompson RC (1990) Inorg Chem 29:4457
69. Bray WC, Livingston RS (1923) J Am Chem Soc 45:1251
70. Clague MJ, Keder NL, Butler A (1993) Inorg Chem 32:4754
71. Clague MJ (1994) PhD dissertation, UC Santa Barbara
72. Butler A, Clague MJ, Meister G (1994) Chem Rev 94:625 .
73. Colpas GJ, Hamstra BJ, Kampf JW, Pecoraro VL (1996) J Am Chem Soc 118:3469
74. Colpas GJ, Hamstra BJ, Kampf JW, Pecoraro VL (1994) J Am Chem Soc 116:3627
75. Tromp MGM, Olafsson G, Krenn BE, Wever R (1990) Biochim Biophys Acta 1040:192
76. Butler A, Carrano CJ (1991) Coord Chem Rev 109:61
77. Neumann R, Assael I (1989) J Am Chem Soc 111:8410
78. Lydon JD, Schwane LM, Thompson RC (1987) Inorg Chem 26:2606
79. Ghiron A F, Thompson RC (1989) Inorg Chem 28:3647
80. Kusthardt U, Hedman, B, Hodgson KO, Hahn R, Vilter H (1993) Febs 329:5
81. Kato T, Ichinose I, Kamoshida A, Hirata Y (1976) Chem Commun 518
82. He H-Y, Faulkner DJ, Shumsky JS, Hong K, Clardy J (1985) J Org Chem 54:2511
83. Carter GT, Rinehart KL, Li LH, Kuentzel SL, Conner JL (1978) Tetra Lett 46:4479
84. Foote CS (1979) In: Wasserman HH and Murray RW (eds) Singlet Oxygen Academic Press, Orlando, FL, p 139

Metal Ions in the Mechanism of Enzyme-Catalysed Phosphate Monoester Hydrolyses

David Gani* and John Wilkie

School of Chemistry and Centre for Biomolecular Science, The Purdie Building,
The University, St. Andrews, Fife KY16 9ST, UK.
E-mail: (David Gani) dg@st-and.ac.uk (John Wilkie) jw5@st-and.ac.uk

Phosphate monoester hydrolysis is of central importance to the biochemistry of living cells. We present an overview of the roles of metal ions in the catalysis of this process in a number of enzymes, drawing together both structural and kinetic information. The enzymes described utilise several different catalytic mechanisms and are active in a wide range of environments. However, kinetic data suggest that a group of enzymes with quite different substrate specificity and preferred environments share a common catalytic mechanism. We also describe the structure and mechanism of two phosphatases that do not require metal ions for activity and compare the roles of specific residues with the bound metal ions of their metal-containing counterparts.

Keywords: Phosphatase, Hydrolysis, Mechanism, Catalysis, Metal Ion.

	Abbreviations	134
1	Introduction	134
2	Phosphate Monoester Hydrolysis	136
2.1	Non-Enzymic Reactions	136
2.2	Enzyme Models and the Role of Metal Ions	137
2.3	Stereochemistry and Mechanism of Phosphate Ester Hydrolysis	138
3	Phosphatase Enzyme Mechanisms	139
3.1	Alkaline Phosphatase	139
3.2	Acid Phosphatase	143
3.3	Purple Acid Phosphatase (PAP)	146
3.4	Protein Tyrosine and Protein Serine, Threonine Phosphatases	149
3.4.1	Protein (phospho)-Tyrosine Phosphatases (PTPs)	150
3.4.2	Protein (phospho)-Serine/(phospho)-Threonine Phosphatases	151
3.4.3	Histidine Protein Phosphatases	159
3.5	5'-Nucleotidase	159
3.6	Inositol Monophosphatase	160
3.7	D-Fructose 1,6-Bisphosphate 1-Phosphatase	168
4	Overview	170
5	References	173

---

* Corresponding Author.

Structure and Bonding, Vol. 89
© Springer Verlag Berlin Heidelberg 1997

Abbreviations

ADP   Adenosine diphosphate
AMP   Adenosine monophosphate
ATP   Adenosine triphosphate
GMP   Guanosine monophosphate
IMP   Inosine monophosphate
M^{2+} Generic divalent metal ion
PAP   Purple Acid Phosphatase
P_i Inorganic phosphate
PP1   Protein Phosphatase 1
PP2A   Protein Phosphatase 2A
PP2B   Protein Phosphatase 2B
PP2C   Protein Phosphatase 2C
PTP   Protein Tyrosine Phosphatase
t-RNA   Transfer RNA

1
Introduction

The transfer of phosphoryl groups from one entity to another is, unarguably, the most important mechanism by which cellular function is orchestrated. Hundreds of metabolic pathways are controlled by the reversible phosphorylation-dephosphorylation of a multitude of different chemical entities and a wide range of different kinase and phosphatase enzymes exist to maintain the correct balance of metabolites. Important examples include the oxidation of glucose to provide energy for the cell, the storage of glucose as glycogen and its reformation, the activation of nucleosides to provide the building blocks for DNA and RNA, the biosynthesis of amino acids and aminoacyl t-RNA molecules to provide the building blocks for proteins and the switching-on and turning-off of many enzymes involved in signal transduction and the control of cellular metabolism. The dynamic control of the opposing processes of glycogen synthesis and its breakdown serve to illustrate how phosphorylation and dephosphorylation can work at a metabolic level. The activity of the key enzymes in each process, glycogen synthetase and phosphorylase, respectively, is controlled by phosphorylation, as are their kinases. In response to adrenaline, mediated via cyclic AMP, both glycogen synthetase and phosphorylase become phosphorylated, inhibiting the former while activating the latter. Simultaneously the kinases for the two enzymes, protein kinase and phosphorylase kinase, are also phosphorylated, activating them further. This serves to amplify the phosphorylation of both glycogen synthetase and phosphorylase. In the absence of cyclic AMP, phosphorylation of the kinases ceases, and they are rapidly dephosphorylated by protein phosphatases, thus suppressing phosphorylation of glycogen synthetase and phosphorylase. The same protein phosphatases also dephosphorylate both glycogen synthetase and phosphorylase, leading to the activation of glycogen synthetase and the inhibition of

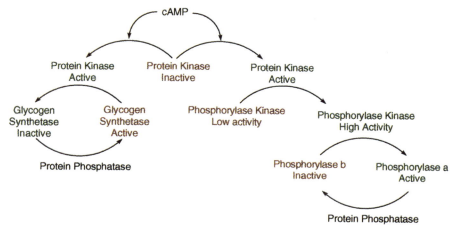

Fig. 1. The Glycogen Synthetase/Phosphorylase cascade. Phosphorylated enzymes are shown in green, dephosphorylated forms in red

phosphorylase. Thus in the absence of the cyclic AMP signal, glycogen synthesis is favoured over its breakdown (see Fig. 1).

While the glycogen synthesis/breakdown story is now well understood, thanks to the pioneering efforts of Krebs and Fischer [1] and many others, other processes are still poorly understood, particularly at a molecular level. The next ten years promises to be full of surprises as the intricacies of neurotransmission, intracellular signal transduction and receptor site-mediated responses are clarified.

Kinase enzymes are responsible for transferring phosphoryl groups to phosphoryl group acceptor moieties within molecules. These sites are usually, but not always, O atoms. The accepting O atom can be part of a wide range of different functional groups, including alcohols, enols, phenols, carboxylates and phosphates. The structures and sizes of the acceptor molecules are diverse. Kinase enzymes exist to phosphorylate small molecules, such as acetate, pyruvate, D-glucose and adenosine, as well as large molecules, such as the serine, threonine and tyrosine residues of a vast array of proteins. Usually, adenosine 5'-triphosphate (ATP) serves as the phosphorylating agent in these reactions, and the O atom of the acceptor molecule (X = OH) directly displaces adenosine 5'-diphosphate (ADP), an excellent leaving group, from the γ-phosphorus atom of ATP (Scheme 1).

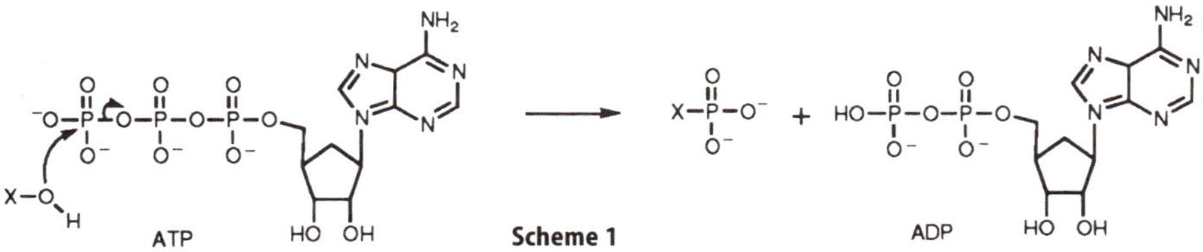

Scheme 1

Phosphatase or phosphohydrolase enzymes are responsible for the removal (hydrolytic cleavage) of phosphoryl groups from an equally diverse range of molecules (Scheme 2). This short chapter reviews our understanding of the action, mechanism and structure of several different phosphatases, some which utilise metal ions, and some which do not. We highlight the chemistry of the hydrolytic process for each system and try to put into context how nature has arrived at different solutions to what is essentially the same chemical problem; the hydrolysis of a phosphate ester. We shall concern ourselves only with monophosphate ester hydrolyses where the leaving group is an alcohol.

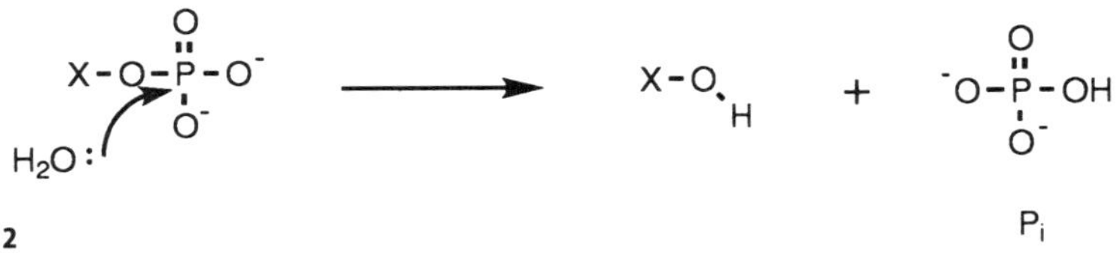

Scheme 2

P_i

2
Phosphate Monoester Hydrolysis

2.1
Non-Enzymic Reactions

In essence, as for the carboxylic esters, there are two fundamental mechanisms by which a phosphate monoester can be hydrolysed. In the first, the C-1 carbon atom of the alcohol can serve as the electrophile, so that the entire phosphate group is replaced by the O atom of a water molecule or hydroxide. In the second, water or hydroxide attacks the phosphorus atom and displaces the alcohol or alkoxide with an intact C-O bond via phosphorus-oxygen bond fission. In principle, these two mechanisms are easily distinguished by performing hydrolyses in ^{18}O-water, since they give different isotopically labelled products, but in practice, acid-catalysed water-phosphate O atom scrambling can complicate analyses (Scheme 3).

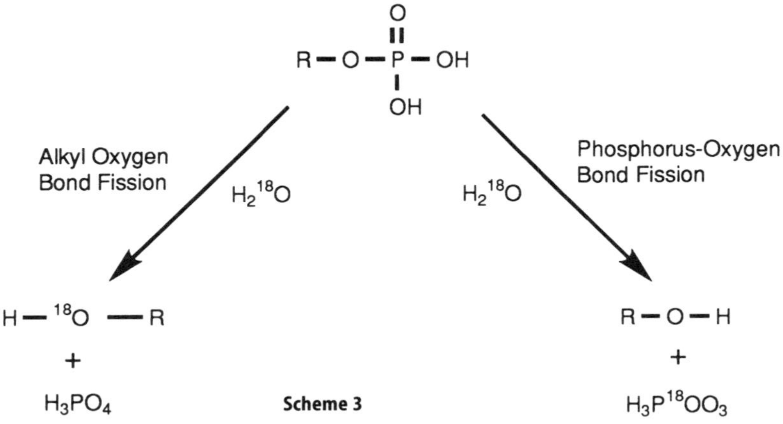

Scheme 3

The alkyl-O fission mechanism is only important at very low pH, where the leaving group is neutral phosphoric acid. For alkyl esters possessing alkyl groups that form stable carbocations, for example benzyl and t-butyl phosphate and α,D-glucose 1-phosphate, the reactions proceed via an S_N1 ionisation process. Conversely, at very low pH, methyl phosphate and primary and secondary alkyl phosphate esters are hydrolysed via two competing processes, S_N2 alkyl-oxygen bond cleavage and phosphorus-oxygen bond cleavage [2]. Above pH 1.5, where the leaving phosphate group possesses at least one negative charge, hydrolyses proceed via phosphorus-oxygen bond cleavage mechanisms. This mode of cleavage is followed by all enzyme-catalysed phosphate monoester hydrolyses reported to date, and thus the phosphorus atom serves as the electrophile.

2.2
Enzyme Models and the Role of Metal Ions

Many phosphatase enzymes employ metal ions as cofactors to lower the activation energy for P-O bond fission. The exact mechanism by which this is achieved depends on the specific enzyme. However, it is known from model studies that metal ions can contribute in three distinct ways. First, the association of a water molecule or an alcohol with a metal ion, for example Mg^{2+}, lowers its pK_a value [3–5]. This is achieved because the conjugate base, hydroxide or alkoxide, can dissipate its charge through a binding interaction with the metal ion; for Mg^{2+} and Mn^{2+} the pK_a values of a bound water molecule are reduced to 11.4 and 10.6, respectively [3]. Since these conjugate bases are better nucleophiles than the parent acids, attack on the electrophilic P atom is enhanced relative to the uncatalysed reaction. Metal ions can also promote reaction by positioning the nucleophile correctly for attack on the P atom. This can be achieved by forming 4-membered cyclic transition states (metallophosphodioxetanes) as shown in Scheme 4. This arrangement provides a further advantage in enhancing the electrophilicity of the P atom through Lewis acid coordination of the metal ion to one of the O atoms that becomes equatorial in the trigonal bipyramidal transition state. Yet further advantage can be obtained by allowing another metal ion to bind to the O atom of the leaving group. Jencks has examined many aspects of metal ion catalysis relevant to phosphatase action in model reactions and reference to the original work is recommended [6–11]. In essence, enzymes appear to fully utilise the catalytic potential of metal ions but, as is highlighted below, there are many subtle differences in the exact way that this is achieved.

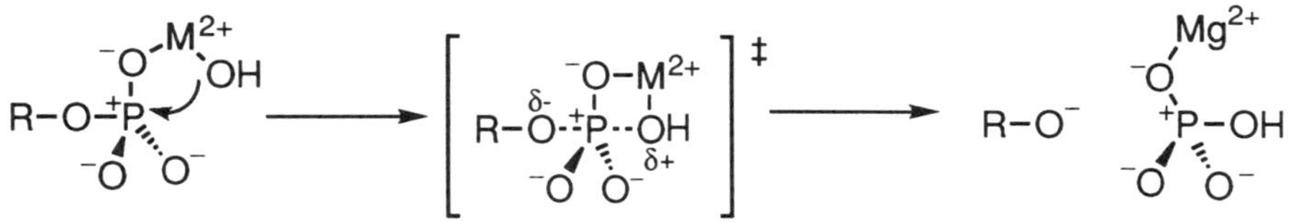

Scheme 4

2.3
Stereochemistry and Mechanism of Phosphate Ester Hydrolysis

Before considering the details of individual phosphatase enzymes, it is worth examining which stereochemical options are available for the transfer of a phosphoryl group from an alcohol to an incoming nucleophile. For phosphatases, following the stereochemical course is far from trivial [12, 13]. The inorganic phosphate product molecule possesses four equivalent O atoms and there are only three stable isotopes of oxygen available which could be used to label the species. Resorting to the use of sulphur as a surrogate for one of the oxygen

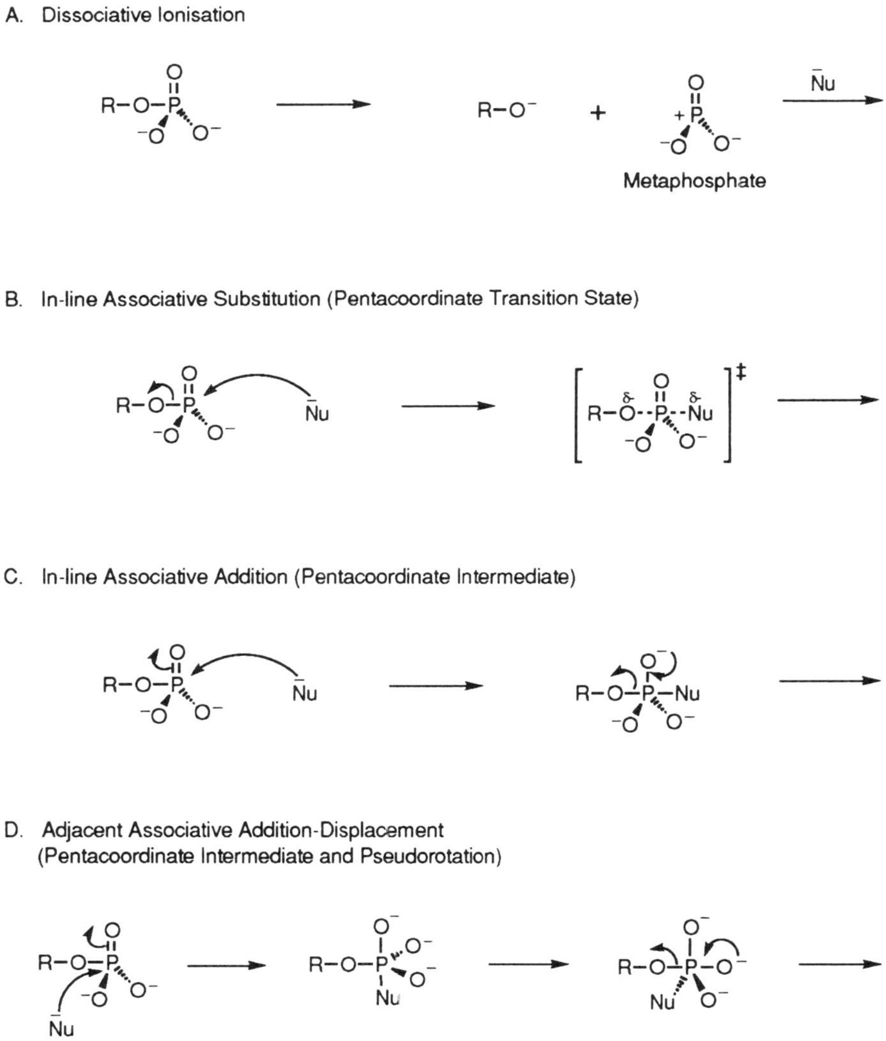

A. Dissociative Ionisation

Metaphosphate

B. In-line Associative Substitution (Pentacoordinate Transition State)

C. In-line Associative Addition (Pentacoordinate Intermediate)

D. Adjacent Associative Addition-Displacement
     (Pentacoordinate Intermediate and Pseudorotation)

Scheme 5

ligands in phosphorothioate substrates, so that the product is a chiral [$^{16}O,^{17}O,^{18}O$]-thiophosphate, superficially resolves the problem, but almost certainly changes the kinetic, and possibly the chemical mechanism of the reaction, see below. Fortunately, the use of phosphorothioate substrates probably does not alter the stereochemical outcome of enzyme-catalysed processes [14].

Of the four possible mechanisms available for phosphoryl transfer in free solution (Scheme 5, A-D) the first (A) should lead to racemisation and two (B and C) should lead to inversion of configuration. Mechanism A is a dissociative ionisation, analogous to an S_N1 process in carbon chemistry, and the nascent trigonal metaphosphate species could, if free, react with a nucleophile on either face to give a racemic product. However, within the environment of an active-site, the out-going negatively charged nucleofuge might shield one face of the metaphosphate species and suppress front face attack such that inversion would be observed. Alternatively, the metaphosphate intermediate may remain associated with the enzyme, leaving only the face from which the leaving group has departed available for nucleophilic attack.

Mechanism B, an in-line associative process, is entirely analogous to an S_N2 displacement in carbon chemistry and would give inversion of configuration. The transition state would be pentacoordinate and the nucleofuge and nucleophile, which occupy the apical positions, would be only partially bonded to the central P atom.

Mechanism C is also an in-line associative process but gives rise to a penta-coordinate stable intermediate with fully formed bonds. The displacement occurs with inversion of configuration, as with mechanism B.

Finally, mechanism D gives rise to retention of configuration via an adjacent associative mechanism. Here the pentacoordinate stable intermediate must pseudorotate (swap ligands about the P atom between apical and equatorial positions) so that the nucleofuge leaves from an apical position [15]. At the present time there is evidence to support the operation of each of the four mechanisms and no compelling reasons to discount any. Clearly, differentiating between mechanisms (A, B and C in Scheme 5) within the confines of the active-site of an enzyme is not possible based on stereochemical information alone.

3
Phosphatase Enzyme Mechanisms

3.1
Alkaline Phosphatase

Alkaline phosphatase is a non-specific phosphomonoesterase which shows maximum activity at pH 9–10, hence its name. The enzyme occurs in both prokaryotes and eukaryotes and is, without question, the most well-studied phosphatase. The enzyme from *E. coli* is a homodimer (M_r = 94,000 Daltons) consisting of 449 amino acid residue subunits and contains two active-sites. Each active-site binds two Zn^{2+} ions and one Mg^{2+} ion [16]. The amino acid sequence of the *E. coli* enzyme shows significant homology (ca. 25–30%) with mammalian alkaline phosphatases and the primary structure of the mammalian enzymes have been fitted to the X-ray crystal structure of the bacterial enzyme. All of the key metal ion-binding interactions in the *E. coli* enzyme are preserved through identical or analogous interactions in the mammalian enzymes, and it is therefore reasonable to believe that the enzymes operate by analogous mechanisms.

Early studies suggested and then showed that the enzyme catalyses a two-stage ping-pong (on-off) phosphate ester hydrolysis reaction [17]. In such mechanisms the phosphoryl group is transferred to a group on the enzyme, which is then subsequently hydrolysed, rather than directly to water (Scheme 6).

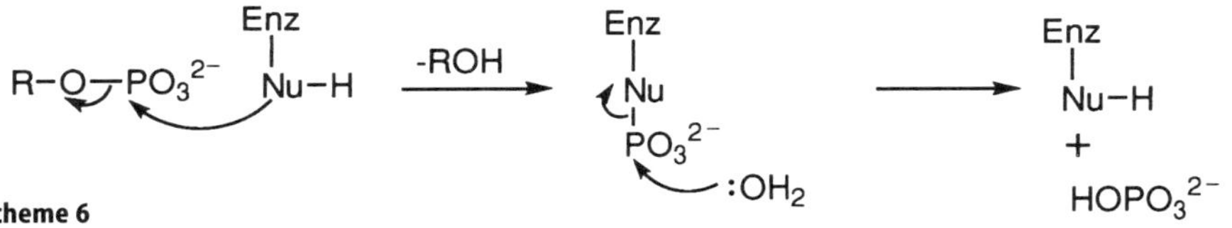

Scheme 6

Evidence for the proposal stemmed from several observations: i) The enzyme gave similar values of k_{cat} for a range of structurally diverse substrates [18] (Note: that these k_{cat} values are suppressed by common late rate-limiting steps, dephosphorylation [at low pH] and P_i dissociation [at high pH]); ii) The enzyme was able to catalyse the transfer of a phosphate group from one alcohol to another (transphosphorylation) [19, 20]; iii) The enzyme was able to catalyse ^{18}O-label exchange from ^{18}O-water to the product, inorganic phosphate (P_i) in the absence of the other product (alcohol); iv) The enzyme displayed burst-phase kinetics for the release of the product alcohol (i.e. an initial rate faster than its steady-state rate of formation and release) at low temperature and v) A phosphoryl enzyme was shown to be formed from free enzyme and P_i at low pH and subsequent studies identified a serine residue in the protein as the phosphorylated moiety [21–23].

Thus, the enzyme displaces the alcohol from the P atom using an active-site serine residue and, at some subsequent time, a water molecule displaces the serine sidechain from the phosphorylated enzyme (E-P) to give an E.Pi complex, and then ultimately the free enzyme and P_i (Scheme 7, path A). Alternatively, in the presence of an alcohol, the serine sidechain of E-P can be displaced to regenerate a new enzyme-substrate complex, which can, of course,

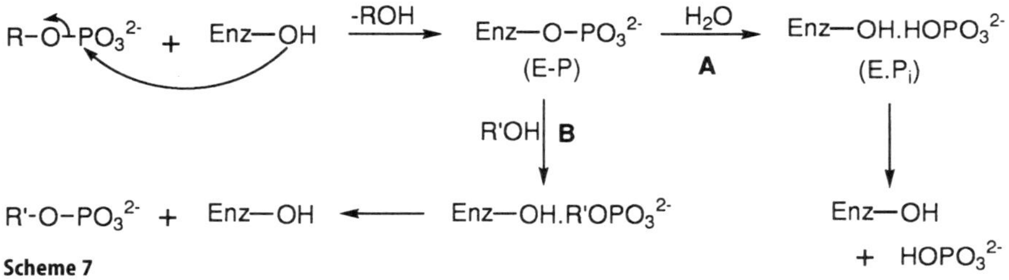

Scheme 7

react to give the alcohol and E-P, or dissociate to give the free enzyme and a new substrate (Scheme 7, path B).

As a consequence of this highly symmetrical mechanism utilised by alkaline phosphatase, the absolute stereochemical configuration at the phosphorus atom is retained upon transphosphorylation, as was demonstrated by Knowles [17]. In the original work, phenyl phosphate, containing a chiral phosphorus centre, underwent transphosphorylation to give (2S)-1-phosphopropan-1,2-diol, which was converted to the cyclic diester form prior to configurational analysis by mass spectrometry (Scheme 8) [Note that since these original studies ^{31}P-NMR spectroscopy techniques have superseded mass spectrometric analysis] [14]. More recent work has shown that a mutant enzyme, in which the active-site serine residue (Ser102) is replaced by cysteine, also gives retentive transphosphorylation. Careful consideration of the transphosphorylation process (Scheme 8) reveals that either two retentions or two inversions would give overall retention. Thus, unfortunately, one can say nothing about the stereochemical course of the individual phosphoryl transfer steps. Nevertheless, some insight has come from crystallographic studies.

In 1991 Kim and Wyckoff reported on an X-ray crystal structure of *E. coli* alkaline phosphatase complexed to P_i at 2.0 Å resolution [24]. This structure

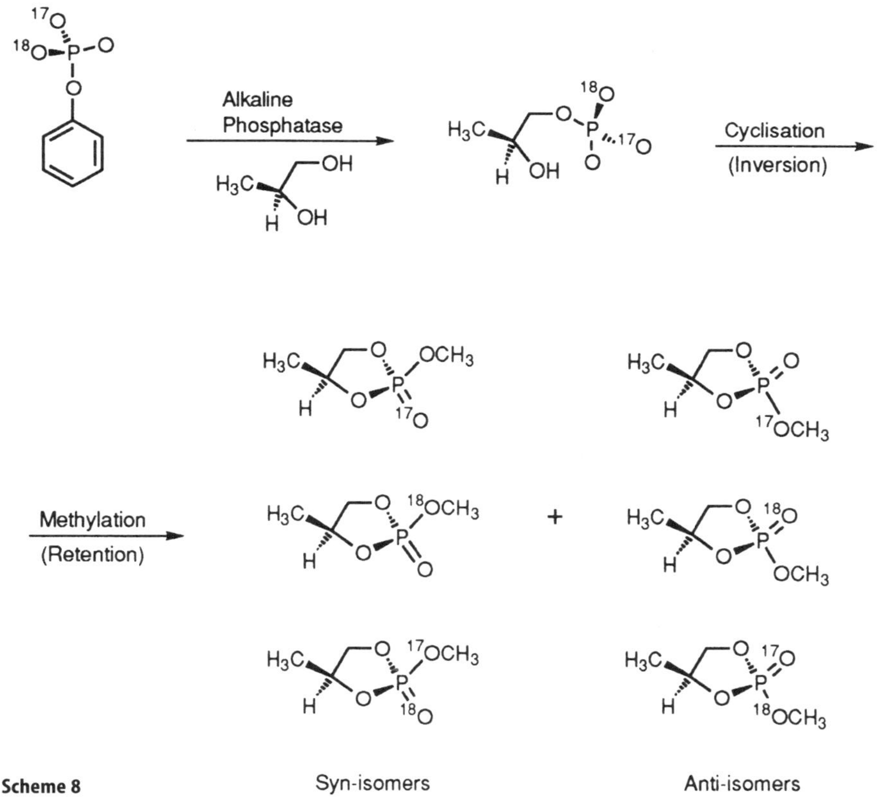

Scheme 8   Syn-isomers   Anti-isomers

was a considerable refinement of an earlier structure with 2.8 Å resolution [25], and allowed a rather thorough examination of the interactions of the protein and the inorganic phosphate molecule with the zinc and magnesium ions (Fig. 2).

Zinc ion number one ($Zn^{2+}1$) is penta-coordinated to the imidazole N atoms of His331 and His412, both carboxylate O atoms of Asp327 and one of the phosphate O atoms. $Zn^{2+}2$ is tetrahedrally coordinated by the imidazole N atom of His370 , one of the carboxylate O atoms of Asp51 and Asp369 and a second of the phosphate O atoms. Thus, both Zn^{2+} ions interact with P_i. The Mg^{2+} ion is octahedrally coordinated with the remaining carboxylate O atom of Asp51, one of the carboxylate O atoms of Glu322, the hydroxy group of Thr155 and three water molecules. Thus, the Mg^{2+} ion does not interact with P_i. The remaining two O atoms of the P_i molecule form a bidentate hydrogen bond with the guanidium group of Arg166 and are thus tightly bound. Each P_i-O atom also forms two additional hydrogen bonds, one with two water molecules (one of which is coordinated to Mg^{2+}), and one with the amide group of Ser102 and a water molecule.

Kim and Wyckoff also obtained a structure for a Cd-substituted enzyme with 2.5 Å resolution which contained a phosphorylated active-site serine residue [24]. Together, the two crystal structures were used to construct a detailed des-

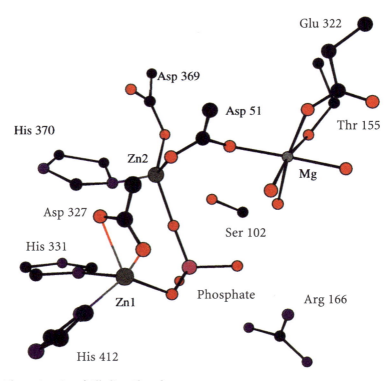

Fig. 2. The active site of Alkaline Phosphatase

cription of the catalytic mechanism (Scheme 9). According to this, the substrate, ROP, first binds to the enzyme such that the phosphate O atom bearing the alkyl group (O1) coordinates to $Zn^{2+}1$. One of the other phosphate O atoms (O2) coordinates to $Zn^{2+}2$ such that the hydroxymethyl O atom of Ser102 is placed diametrically opposite to the alcohol O atom of the leaving group. The process of phosphoryl transfer is enhanced by the hydroxymethyl O atom coordinating to $Zn^{2+}2$. This action causes the phosphate O2 atom to bridge the two Zn^{2+} ions and allows the hydroxy group of Ser102 to deprotonate and attack the P atom. The result is the formation of a pentavalent phosphorus intermediate or alternatively, transition state, which then collapses to an inverted tetrahedron through the departure of the $Zn^{2+}1$-chelated alkoxide O atom (O1) (Scheme 9). Two important mechanistic features of the phosphoryl transfer reaction are defined by this sequences of events. First, the transfer to Ser102 occurs with inversion of configuration and; second, either of the two in-line associative mechanisms discussed above (Scheme 5, B or C) could operate.

Following the cleavage of the O1-P bond, the alkoxide becomes protonated, leaves the coordination sphere of $Zn^{2+}1$, and then dissociates from the active-site. A solvent-derived water molecule then enters the active-site, chelates to $Zn^{2+}1$, deprotonates and then attacks the phosphoryl serine moiety through an in-line displacement . Ultimately, through the reverse sequence of steps to those described above, the non-covalent enzyme-P_i complex is formed and P_i dissociates from the active-site to complete the catalytic cycle (Scheme 9).

3.2
Acid Phosphatase

Very much less is known about acid phosphatases, a family of non-specific enzymes of molecular weight 40,000 – 60,000 Da. The enzymes have been isolated from a variety of sources including yeast, human lysosomes and human prostate as well as plants and bacteria. These enzymes have an optimum pH of 2.5 [26] and do not require metal ion cofactors for catalysis. The enzymes from different sources show limited amino acid sequence homology, although certain regions of the proteins are highly conserved, in particular, a motif near the N-terminal which includes Arg-His-Gly-Xaa-Arg-Yaa-Pro (positions 16–22 in the *E. coli* enzyme) [27]. The enzymes display many characteristics indicative of a ping-pong phosphorylated enzyme intermediate mechanism. These include: i) the formation of a phosphoprotein containing a covalently modified histidine residue [26]; ii) burst-phase kinetics for the release of the product alcohol; iii) transphosphorylation activity [28], and iv) an ability to catalyse ^{18}O-label exchange from the solvent into P_i [29]. The stereochemical course of the transphosphorylation reaction catalysed by the enzyme was determined in an analogous manner to that for alkaline phosphatase, and overall retention of configuration for the two transfer steps was observed [28]. Again it must be stressed that the result provides no information on the stereochemical course of the individual steps.

Recent studies on the *E. coli* enzyme have tested the role of the conserved residues between different species. The alterations of Arg16 to alanine or His17 to

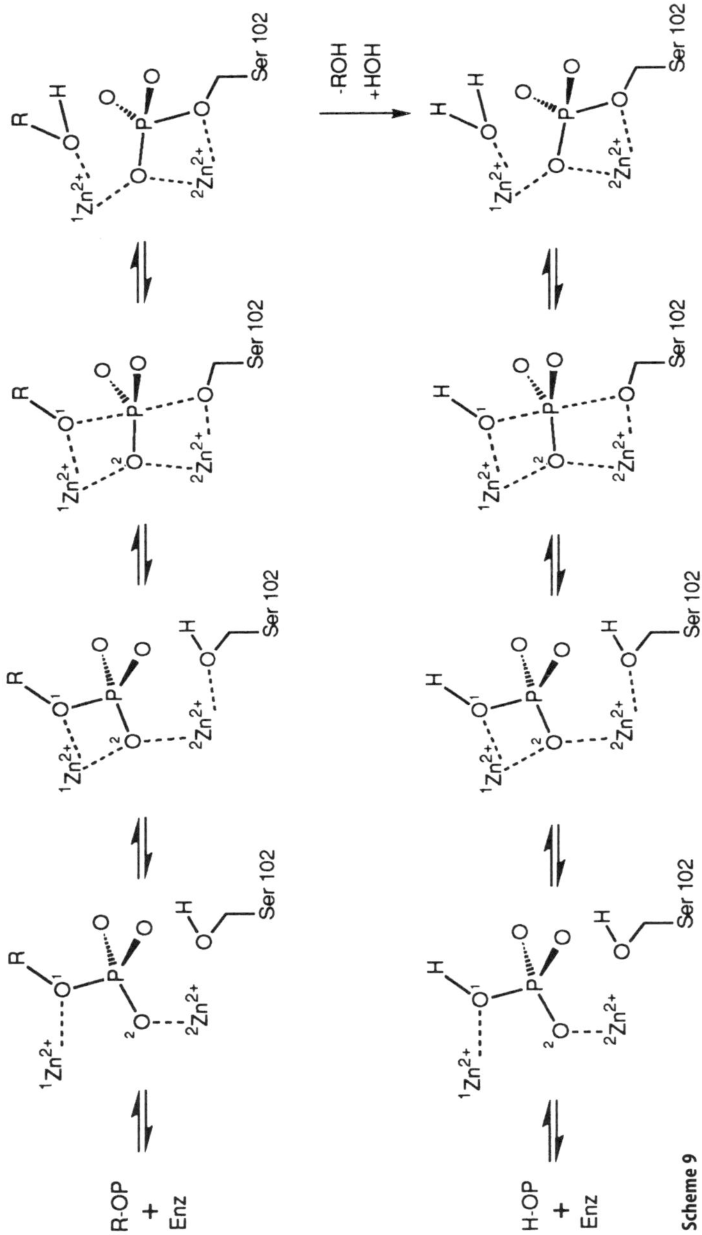

Scheme 9

asparagine gave inactive proteins. By analogy to the role played by the equivalent residues in phosphoglycerate mutase it is believed that Arg16 interacts with both His17 and His303 at the active-site and that these histidine residues serve as the phosphoryl acceptor moiety and as a proton donor respectively (Scheme 10) [30]. The alteration of His303 to an alanine residue gave an enzyme with very low activity, in accord with its proposed role, as did the alteration of the putative substrate binding residue, Arg92 (to alanine). The alteration of Asp304 to Ala changed the rate-limiting step from the hydrolysis of the phosphoenzyme intermediate in the wild-type and His303Ala mutant to formation of the phosphoenzyme intermediate. Thus, Asp304 rather than His303 may be involved in protonating the departing O atom of the alcohol. These results indicate how difficult it is to rationalise the effects of specific mutations.

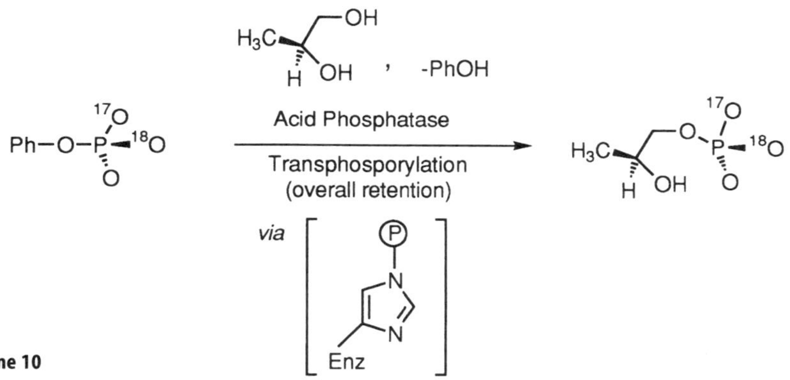

H_3C —OH, H, OH , -PhOH

$Ph-O-P \overset{17O}{\underset{O}{}} {}^{18}O$

$\xrightarrow[\text{Transphosporylation (overall retention)}]{\text{Acid Phosphatase}}$

H_3C, O-$P \overset{17O}{} {}^{18}O$, H, OH, O

via

$$\left[\overset{\textcircled{P}}{\underset{Enz}{N \diagup N}} \right]$$

Scheme 10

   Crystal structures have been obtained for rat prostatic acid phosphatase, which shares the same catalytic motif, in the presence of vanadate [31] and tartarate [32]. Though there is no direct sequence homology between this enzyme and the *E.Coli* acid phosphatase, other than the catalytic motif, prostatic acid phosphatase contains residues in its active-site equivalent to those described above. The vanadate-inhibited structure [31] is clearly an analogue of the pentavalent intermediate bound to the histidine nucleophile, His12 (Fig. 3). Two of the vanadate O atoms form hydrogen bonds with active-site arginine residues, Arg15 and Arg79, which are analogous to Arg20 and Arg92, respectively, in *E.Coli* acid phosphatase. A short loop, comprising residues 253 to 271, does not appear in this crystal structure, but comparison with the tartarate-inhibited structure [32] suggests that the sidechain of His257 and backbone amide of Asp258 may both donate hydrogen bonds to a further vanadate oxygen. The remaining vanadate oxygen, which may be presumed to be analogous to the O atom which dissociates from the substrate because it lies opposite the nucleophile, does not form a hydrogen bond with any enzyme residue. However, the sidechain of Asp258 forms a strong hydrogen bond with one of the carboxylates of tartaric acid and may, with only a small change in sidechain position, form a hydrogen bond with either the departing O atom, protonating it and facilitating its departure, or, depending on its pK_a, accept a hydrogen bond from the nucleo-

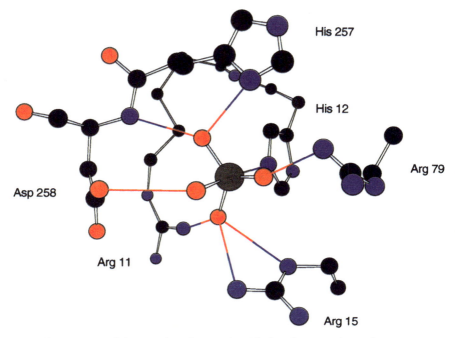

Fig. 3. The structure of the complex of prostatic acid phosphatase and vanadate

philic water molecule thereby activating it and catalysing the hydrolysis of the phosphoro-enzyme intermediate. In the tartarate-inhibited structure, the other carboxylate oxygen of Asp258 forms a hydrogen bond with the second terminal nitrogen of Arg11.

3.3
Purple Acid Phosphatase (PAP)

Purple acid phosphatases occur in bacteria, plants and animals and hydrolyse aryl phosphate monoesters, phosphoric anhydrides and the phosphoserine residues of phosphoproteins. They are characterised by low pH-optima, their insensitivity to tartrate inhibition and by their purple colour. The most extensively studied examples are those isolated from porcine uterus (uteroferrin) and from bovine spleen. The enzymes are both monomeric glycoproteins of about 35,000 Daltons which show ~ 90% amino acid sequence homology. Roles for the mammalian enzymes in the phagocytosis of aged erythrocytes and in the resorption of bone have been suggested. It is also thought that PAPs might serve as protein phosphatases because phosphorylated proteins containing phosphorylated serine residues do serve as substrates for the enzymes.

Purple acid phosphatases contain two Fe ions at the active-site. The oxidised form contains two antiferromagnetically coupled Fe^{3+} ions, is purple in colour and inactive. Reduction gives the active $Fe^{3+} \cdot Fe^{2+}$ form which is pink [33]. Interestingly, purple acid phosphatase from kidney bean contains Fe^{3+} and Zn^{2+} at the active-site and is a homodimer of 111 kDa linked by a disulfide bridge [34]. The Zn^{2+} ion can be replaced by Fe^{2+} to give a protein that is almost spectroscopically and kinetically identical to the mammalian PAPs [35].

Although little active-site structural information has been available until very recently, see below, it was believed that PAPs operated via a phosphorylated enzyme intermediate [36]. Evidence to support such a mechanism stemmed from the observation of i) burst-phase kinetics, ii) transphosphorylation activity and iii) the retention of ^{32}P-label by the enzyme after exposure to the substrate $[\gamma\text{-}^{32}P]$-ATP. However, it was subsequently shown that the enzyme catalyses the transfer of the γ-phosphorothio group of γ-chirally labelled $2',3'$-methoxylidine-γ-phosphorothio-ATP to water with inversion of configuration at the phosphorus atom (Scheme 11) [37]. Although the leaving anionic diphosphate group in this ATP analogue differs considerably from the alkoxy leaving group of monophosphate ester substrates, the most likely implication of this result is that the hydrolytic mechanism involves the direct displacement of the leaving group from the terminal P atom by water in a single step. This interpretation is also supported by the results of electron spin echo envelope modulation (ESEEM) and electron nuclear double resonance (ENDOR) studies which indicated that water molecules were directly coordinated to the metal ions [38].

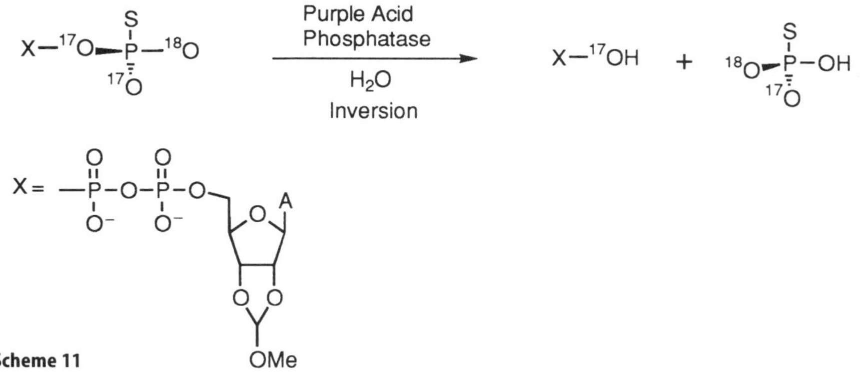

Scheme 11

In 1995 an X-ray crystal structure [34] of the kidney bean enzyme was reported at a resolution of 2.9 Å (Fig. 4). Interestingly, the Zn^{2+} ion is bound in an identical complex to one of the two metal ions of PP1 and PP2B (see following section). It is coordinated by His286, His323, the amide oxygen of Asn201 and by Asp164, which forms a monodentate bridge between the two metal ions. The Fe^{3+} ion is coordinated by Tyr167, His325 and Asp135 in addition to the bridging O atom of Asp164. The moderate resolution of the X-ray structure did not allow the placing of any water molecules. However, subsequent structures containing phosphate and tungstate showed that these species bind to each of the two metals simultaneously through two different O atoms [39]. This arrange-

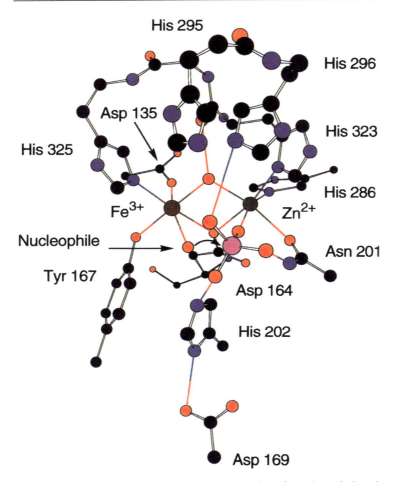

His 295
His 296
His 323
Asp 135
His 325
His 286
Fe^{3+}
Zn^{2+}
Nucleophile
Asn 201
Tyr 167
Asp 164
His 202
Asp 169

Fig. 4. The structure of the proposed reactant complex of purple acid phosphatase and phosphate. The nucleophilic water molecule is bound to the Fe^{3+} ion

ment of ligands gives near-perfect octahedral geometry for the Fe^{3+} ion and a distorted octahedron for the Zn^{2+} and leaves a vacant site for an μ-bridging water. The two non-coordinated phosphate oxygen atoms form hydrogen bonds with His202 and His296, the latter, along with His295, being moved by about 1 Å compared with its position in the phosphate-free enzyme.

It is thought that the enzyme-phosphate complex observed in the crystal structure is the product-inhibited form of the enzyme, and that the two metal ion-binding phosphate O atoms are replaced by water ligands in the free enzyme [39]. In a plausible ground-state for the catalytic mechanism, the Fe^{3+}-bound water has a lowered pK_a and serves as the nucleophile, while the Zn^{2+}-bound water is displaced by an O atom of the substrate phosphate group. Access to the active-site is such that the substrate phosphate could bind with the ester-bridging oxygen atom positioned as for any of the non-coordinating phosphate O atoms in the crystal structure of the E.P_i complex. However, in accord with the observed inversion of stereochemistry at the P atom during the course of the reaction, it is reasonable to expect the substrate to bind such that the O atom of the alcohol leaving group lies diametrically opposite the Fe^{3+}-bound nucleophile. A small amount of movement of the protein backbone would allow His295 to form a hydrogen bond with one of the non-chelating phosphate O atoms, displacing His296 which, in turn, could form a hydrogen bond with the leaving alcoholate O atom, such that it is protonated as it departs. This mechanism, consistent with the kinetically determined roles of the metal ions in binding to, and activating both water and the substrate, is shown in Scheme 12.

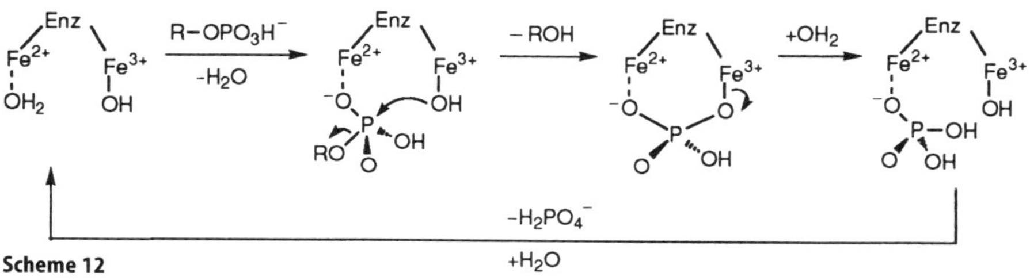

Scheme 12

The kidney bean purple acid phosphatase shows maximum activity at pH 5.9 with α-napthyl phosphate as substrate, and the bell shape of the pH/activity relationship is consistent with pK_a values of 4.8 and 6.9 in the active complex [39]. Thus the enzyme requires one functional group to be protonated and another deprotonated for full activity. The first pK_a value is consistent with the deprotonation of the Fe^{3+}-bound water, while a pK_a of 6.9 could correspond either to the deprotonation of the imidazolium sidechain of an enzyme His residue, or with the second ionisation of the phosphate group. Deprotonation of the phosphate, to give the dianion, would be expected to reduce the electrophilicity of the phosphorus atom and make it less susceptible to nucleophilic attack, while deprotonation of an essential histidine, such as His296, may prevent protonation of the leaving alkoxide group and thus retard its departure.

3.4
Protein Tyrosine and Protein Serine, Threonine Phosphatases

Protein phosphatases are a diverse group of enzymes responsible for the dephosphorylation of a range of phosphoproteins. Many are involved in the regulatory control of cellular processes as diverse as cell growth and proliferation, protein, cholesterol and fatty acid biosynthesis and glycolysis/gluconeogenesis. The protein phosphatases can be categorised into two large groups, those that dephosphorylate phosphotyrosine residues within proteins and those that dephosphorylate phosphoserine or phosphothreonine residues within proteins. The possibility of another group of enzymes that serve as protein phosphohistidine phosphatases is also considered. However, it is likely that at least some of these enzymes are actually protein Ser/Thr phosphatases which display activity towards phosphohistidine-containing substrates.

3.4.1
Protein (phospho)-Tyrosine Phosphatases (PTPs)

Essentially, the phosphotyrosyl protein phosphatases can be divided into two groups, the low-molecular weight and the high-molecular weight forms. The high-molecular weight forms can be further classified by whether or not they are associated with, and modulated by, receptors. The receptor-modulated PTPs have received particular attention in recent years and this important group of enzymes are now known to be involved in the regulation of cell growth, proliferation and differentiation [40-42].

In common with the acid phosphatases and in contrast with all the other phosphatases discussed so far, protein tyrosine phosphatases (PTPs) do not bind metal ions in their active-sites. This is true of both the high- and low-molecular weight PTPs regardless of whether they are eukaryotic, prokaryotic or viral in origin. In common with the acid phosphatases, PTPs are most active at a pH at which phosphate ester substrates cannot exist as the dianion, thus the electrophilicity of the P atom cannot be further reduced by an extra negative charge. However, the electron-withdrawing nature of the phenyl group enhances the electrophilicity of PTP substrates over their alkyl phosphate counterparts and it may be that nature has reserved the use of metal ions only for the least reactive of phosphatase substrates.

High-molecular weight phosphotyrosyl protein phosphatases, whether associated with receptors or not, share a conserved catalytic domain of 240 amino acid residues [40], incorporating the structural motif characteristic of the class, namely (I/V)HCxAGxGR(S/T)G. On the other hand, the low-molecular weight cytoplasmic phosphotyrosyl protein phosphatases, thought to be important in the intracellular phosphoprotein dephosphorylation, do not show sequence homology to the larger enzymes, but share the same structural motif, CxxxxxR, at the active-site.

The kinetics and mechanism of the low-molecular weight forms have been studied extensively. Early work indicated that the enzyme catalyses transphosphorylation and a stereochemical study showed that the reaction occurred with overall retention of configuration [43]. Further work with the bovine heart enzyme demonstrated the existence of a phosphorylated enzyme intermediate which was later shown to involve the thiomethyl group of Cys12 [44, 45].

In 1994 the area received a massive boost when subsequent work with bovine heart PTP and bovine liver PTP (and also with the high-molecular weight human PTP 1B, see below) resulted in the publication of X-ray crystal structures with 2.2 and 2.1 Å resolution [46-48]. These showed that the motif CxxxxxRS lies within the active-site of the PTP, and wraps around the phosphate group such that five out of six amide nitrogen atoms form hydrogen bonds with the non-leaving oxygen atoms of the phosphate group (see Fig. 5a). The nucleophile, Cys12, is ready for an inline displacement of the leaving group, [46, 47] in accord with the observed overall retentive stereochemical course of the reaction [43].

The conserved Arg18 forms bidentate hydrogen bonds with two of the non-bridging phosphate oxygen atoms but there are no other positively charged residues within the active-site. Ser19 forms hydrogen bonds with the sidechain of Cys12 and may help to activate it, while Asp129 is well placed to form a hydrogen bond with the oxygen of the leaving group. Mutation of Asp129 has shown that loss of this hydrogen bond greatly reduces k_{cat}, but has little effect on K_m for p-nitrophenol phosphate [49, 50]. It also changes the rate limiting step from breakdown of the phosphoro-enzyme intermediate to its formation. There are a pair of tyrosine residues, Tyr131 and Tyr132, close to the active-site that may be able to bind the phenyl ring of the tyrosine residue of the substrate.

Human PTP1B has also been crystallised and the structure shows little correspondence with that of the low-molecular weight PTP except for the consensus sequence (Fig. 5b) [51, 52]. Cys215 acts as the nucleophile and all six of the amide nitrogen atoms donate hydrogen bonds to the non-bridging phosphate O atoms. Arg221 fulfils an identical role to that of Arg18 of the low-molecular weight PTP, while Ser222 forms a hydrogen bond with the sidechain of Cys215. Superposition of the two active-site regions shows that Phe182 (PTP1B) occupies a similar position to that occupied by Tyr131 of the low-molecular weight PTP and can form a π-complex along with Tyr46 (Fig. 5b). This latter residue, Tyr46, has no counterpart in the low-molecular weight structure. In addition, Asp48 forms a hydrogen bond with both the amide nitrogen of the phospho-tyrosyl residue and the following residue in the substrate. The target sequence for the PTPase (i.e. EGF receptor; DADEpYLIPQQG) is largely hydrophilic in nature in agreement with the requirements of the equally hydrophilic environment of the active-site. The enzyme forms a number of binding interactions with the peptidic fragment of the substrate, as required for recognition and specificity, and the complex appears to be set up, primarily, to catalyse the first step in the reaction, namely the formation of the phosphoroenzyme intermediate. After the dephosphorylated peptide has been released, Asp129 may activate a water molecule to attack the P atom and regenerate the free thiol of Cys215 and P_i.

3.4.2
Protein (phospho)-Serine/(phospho)-Threonine Phosphatases

The phosphoserine-phosphothreonine protein phosphatases are a ubiquitous group of enzymes which constitute the catalytic domains of several multiprotein complexes. The enzymes are classified by their sensitivities to various protein inhibitors (inhibitors 1 and 2) and by their cofactor dependencies [53]. Phosphoprotein phosphatase 1 (PP1) contains a catalytic domain of $\sim 35,000$ Da and is inhibited by inhibitor-1 and inhibitor-2. The type 2 enzymes are not significantly affected by inhibitors 1 and 2, but the catalytic domains of PP2A ($\sim 35,000$ Da) and PP2B, calcineurin, (60,000 Da) show significant sequence homology to PP1. PP2C (46,000 Da) appears not to be related [54] PP1 and PP2A did not appear to require divalent metal ions for activity [53], but it is now known that two divalent Mn^{2+} cations are tightly bound at the active-site. PP2B requires Ca^{2+} and binds calmodulin while PP2C requires Mg^{2+} for activity [53].

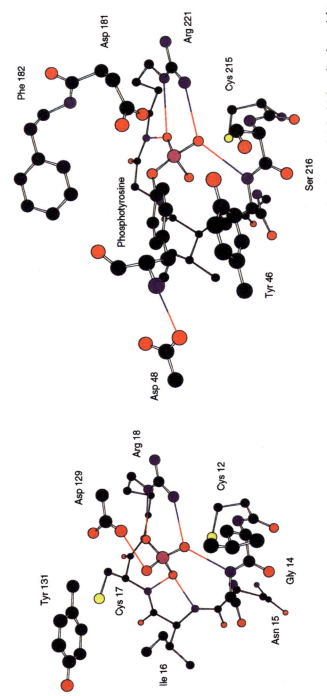

Fig. 5. The structures of the complexes of, left; low molecular weight Bovine Tyrosine Phosphatase and phosphate, and right; high molecular weight Human Tyrosine Phosphatase and a phosphotyrosyl inhibitor

Catalysis by PP1, PP2A and to a lesser extent, PP2B is inhibited by the tumour promoter okadaic acid (**1**) but PP2C is insensitive [53] PP1 and PP2A are also inhibited by microcystin (**2**) and nodularin (**3**) [55].

Sequence comparisons for a wide range of Ser/Thr phosphatases has shown the presence of a consistent motif located in three parts of the polypeptide, namely DxH......GDxxD......GNH(D/E). A similar motif is also found in related enzymes, such as the purple acid phosphatases (PAPs) and the 5′ nucleotidases, (Table 1) and there is no longer any doubt that several structural aspects of the

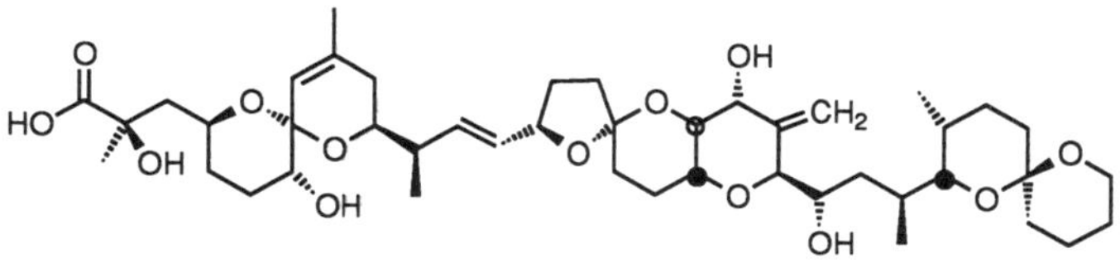

Okadaic Acid
(**1**)

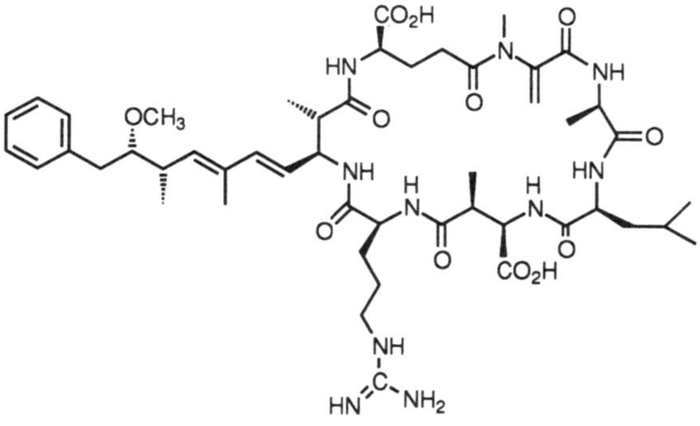

Microcystin-LR
(**2**)

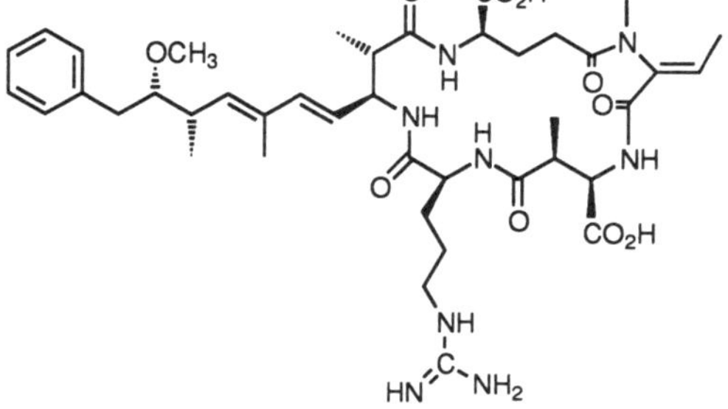

Nodularin
(**3**)

Table 1. Sequence alignment of the Ser/Thr phosphatase and related sequences

λPP	(14)	niwvvGDlHGcytnl	.14.	llisvGDlv-DRGaenve	.9.	fravrGNHEqmmid (82)
PP1	(55)	pikicGDiHGqytdl	.13.	nylflGDyv-DRGkqsle	.15.	fflirGNHEcasin (128)
PP2A	(48)	pvtvcGDiHGqfydl	.13.	nylfmGDyv-DRGyysve	.15.	ltilrGNHEsrqit (121)
PP2B	(79)	pitvcGDiHGqffdl	.13.	rylflGDvy-DRGyfsie	.15.	lsllrGNHEcrhlt (152)
PAP	(129)	tfgliGDlgqsfdsn	.14.	tvlfvGDlsyaDRYpnhdn	.18.	wiwtaGNHEiefap (208)
H5	(30)	tilhtNDvHSrleqt	.34.	llldaGDqyqgtiwftvyk	.13.	damalGNHEfdngv (124)
V5	(39)	tvlhtNDhHGrfwqn	.28.	lllsgGDintgvpesdlqd	.13.	damalGNHEfdnpl (127)

λpp, λ-protein phosphatase [56]; PP1, Green Algae Protein phosphatase 1 [74]; PP2A, Mouse-ear Cress Protein phosphatase 2a [75]; PP2B, Calcineurin from Fruit Fly [76]; PAP, Kidney Bean Purple Acid Phosphatase [77]; H5, Human 5′-Nucleotidase [72]; V5, *Vibrio Parahaemolyticus* 5′-Nucleotidase [73].

Capitals denote conserved metal binding residues, constituting the characteristic Ser/Thr phosphatase motif. Numbers in parentheses denote the position in the sequence of the first and last residues given. Numbers not in parentheses give the number of intervening residues.

different groups are similar. Whether these translate to mechanistic similarities is an interesting question and a detailed comparison of the Ser/Thr phosphatases with the PAPs group is made below.

The smaller but related enzyme, the λ-phosphatase from the λ-bacteriophage, which requires Ni^{2+} or Mn^{2+} for activity, shows considerable sequence homology to PP1 and PP2A [56]. To date this enzyme has been the subject of the most extensive mechanistic studies for the group, in an area where little has yet been done. Even though at the time of the work there was no guiding 3-D structural information available for the system, several of the residues expected to be involved in catalysis were altered in order to assess their roles [56]. These studies indicated that Asp20, His22, Asp49 and His76 are all required for activity, while mutation of Asp52 to Asn reduced k_{cat} 40-fold. The alteration of Glu59 and Glu77 to Gln reduced k_{cat} 8- and 60-fold, respectively, although none of the mutations have any effect on the value of K_m for *p*-nitrophenyl phosphate. Mutation of Arg53 to Ala reduced k_{cat} 50-fold and increased the K_m value by a factor of 20 in the presence of Ni^{2+}, but had no effect on K_m in the presence of Mn^{2+}. Mutation of Arg73 to Ala caused a similar, but smaller effect.

Recently, crystal structures for both PP1 and PP2B have been published [55, 57]. The PP1 structure is of the complex with the inhibitor microcystin [55] which, while not of direct relevance to understanding the mechanism of the Ser/Thr phosphatase, provided a very convincing structural model to account for the extreme potency of microcystin and nodularin which could be used for designing synthetic inhibitors [58]. Furthermore, the crystallographic information for the metal ion-binding sites immediately confirmed the importance of the residues in the conserved motif and allowed the kinetic results of the alteration of specific residues in the λ-phosphatase to be rationalised, as is outlined below.

Asp64 and His66 coordinate one of the metal ions (Mn^{2+}1 in Fig. 6), while the other metal ion, Mn^{2+}2, binds to Asn124 and His248. Asp92 coordinates to both metal ions, leaving only Asp95, His125 and Glu126 from the consensus sequence. Glu126 lies some distance from the metal ion binding site and active-site, but

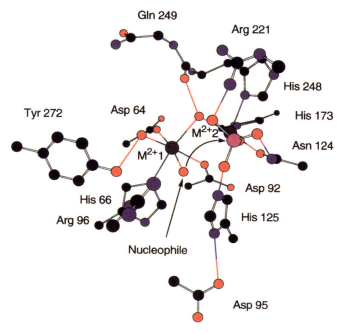

Fig. 6. The structure of the proposed reactant complex of protein phosphatase 1 and phosphate. The nucleophilic water molecule is coordinated to metal ion 1 ($M^{2+}1$)

Asp95 forms a hydrogen bond with the δ-nitrogen of His125, thus raising the pK_a value for the deprotonation of the ε-nitrogen. The environment of $Mn^{2+}2$ is identical to that of the Zn^{2+} ion of purple acid phosphatase, while the binding site for $Mn^{2+}1$ differs from the Fe^{3+} site of purple acid phosphatase only in that His66 and a water molecule substitute for Tyr167 and His325 (see Fig. 4). In all, there are three water molecules coordinated to the metal ions, one of which has been described above. A second forms an μ-bridge between the two metal ions while the third coordinates to $Mn^{2+}2$, forms a hydrogen bond to both His125 and Asn124 and thus occupies a site equivalent to that proposed for the phosphate binding site of PAP.

Despite this close similarity with the active-site of PAP, there are some important mechanistic differences between the protein phosphatases and the purple acid phosphatase. PP2B does not catalyse the oxygen ligand exchange with inorganic phosphate in the presence of divalent metal ions either in the presence or absence of an alcohol product [59]. Moreover, PP2B, in contrast to PAP, does not catalyse transphosphorylation [60] and no trace of a phosphoro-enzyme intermediate has been found. These findings all point towards the rapid release of inorganic phosphate from the product complex, and indeed, PP2B displays V_{max} values that vary significantly for different substrates [61]. It has been suggested that the products are released in a random order [61] since both P_i and alcohols act as competitive inhibitors. This contrasts with the situation for the non-specific alkaline and acid phosphatases, where phosphate release is rate-determining, so many substrates show similar values for V_{max} and phosphate is always released last. From this analysis it would appear that in hydrolysing phosphate esters, PP2B and, by analogy, PP1 utilise a mechanism involving direct attack on phosphorus by either hydroxide or water. It is believed that a similar mechanism is also utilised by both purple acid phosphatase and 5'-nucleotidase[39]. The validity of this proposal will become clearer when more kinetic data has been accumulated and interpreted and the stereochemical course of the reaction has been determined.

The pH dependence of PP2B varies with the metal ions in the active-site; whereas PAP shows maximum activity at pH 5.9 [39]. PP2B is maximally active at pH 7.0 in the presence of Mn^{2+} and pH 8.2 in the presence of Mg^{2+} [61]. A more detailed analysis of the pH profile for PP2B containing Mn^{2+} showed that V_{max} increased with pH to reach a plateau at and above pH 7.0, while the value of V_{max}/K_m declined at a pH above 8.0. While it should be remembered that such pH-profile analyses give only molecular pK_a values, the first deprotonation shows a pK_a value of 6.45. This value could be consistent with the ionisation of an Mn^{2+}-bound water molecule in the E.S complex. Indeed the process is analogous to the activation of the Fe^{3+}-bound water nucleophile in purple acid phosphatase described earlier. The second deprotonation, which produces a decrease in V/K only, is more difficult to assign as it lies close to the pK_a values of a number of enzyme residues, including that of histidine, and also to the pK_a value for the protonation of the phosphate dianion.

There is a modest solvent deuterium isotope effect of 1.35 that is expressed in V/K but not V_{max}. Thus the rate-limiting step does not involve proton transfer, and a slow step, which follows the isotopically sensitive step, brings the faster step into equilibrium [61]. The pH-activity profile for PP2B indicates that the K_m value for the substrate, p-nitrophenyl phosphate rises with increasing pH from 3 mM at pH 5.8 to 110 mM at pH 8.8, [61] suggesting that the enzyme prefers to bind the monoanion rather than the dianion.

A number of mechanisms have been suggested for both purple acid phosphatase [39] and the various Ser/Thr protein phosphatases [57, 59] that differ to a greater or lesser extent. However, it seems likely, in view of the structural similarity between the active-sites, that these enzyme share a common mechanism.

The invertive stereochemical course of the purple acid phosphatase and the lack of transphosphorylation, oxygen exchange and phosphoro-enzyme intermediate for PP2B all point to the mechanism involving a direct attack of water or a hydroxide nucleophile on the phosphorus atom. Kinetic data suggest that this nucleophile should reside on the Fe^{3+} ion of PAP, [62], although Griffith et al suggest the nucleophile could be the μ-oxo bridge water molecule [57] of PP2B. Critical to this latter interpretation is the role of His125 in PP1, the equivalent of which is essential for the activity of the λ-phosphatase. If the μ-oxo bridge water functioned as the nucleophile, His125 would be well placed to form a hydrogen bond with the departing oxygen group, thus facilitating its departure and His125 would therefore contribute to both substrate binding and catalysis. It must, however, be pointed out that a water molecule coordinated to two metal ions is bound significantly more tightly than if it were coordinated to a single ion. Hence it would be much less likely to be able to function as a nucleophile. Moreover, computational work has shown that the energy barrier height for phosphate oxygen exchange, and, by analogy, phosphate hydrolysis, declines dramatically with increasing phosphate protonation [63].

The role of His125 and its analogues in the other phosphatases is something of an enigma. It is required for catalysis, but not apparently for binding, as has been shown by the λ-phosphatase mutation studies [56]. Its pK_a value is increased by the presence of Asp95, yet the ionisation in the pH profiles that most closely fits the expected properties of His125 is associated with a change in V/K rather than in V_{max}. That the His to Asn λ-phosphatase mutant shows no change in K_m compared with the wild type may be misleading, as Asn is very similar to His. Both possess a δ-nitrogen and can donate a hydrogen bond to the aspartate and, although Asn lacks the ε-nitrogen, the δ-oxygen may form a hydrogen bond to a water molecule, positioning it to donate a hydrogen bond in place of the His ε-nitrogen. Thus there might be little alteration to the K_m value, but any subsequent process that required a proton transfer, either fully or partially, from this nitrogen would be disrupted. This assumes, of course, that the activity of the His125Asn mutant was not due to low levels of contaminating wild-type enzyme.

A plausible mechanism (Scheme 13) which fits just about all the known kinetic data is one in which the phosphate group of the substrate displaces a water molecule from either the Zn^{2+} ion of PAP or the $M^{2+}2$ ion of the protein phosphatases, such that one of the non-bridging phosphate O atoms forms a hydrogen bond with His125 (His202 in PAP). The location of Asp95 is such that the pK_a values of the His ε-nitrogen and the phosphate ester group are closely matched. Thus their close proximity would result in a considerable degree of proton transfer to the phosphate O atom and this may be the rapid, pre-rate-limiting step that is observed as the solvent equilibrium isotope effect. The $M^{2+}1$-bound hydroxyl would then be able to attack the electrophilic phosphorus atom to form the pentavalent intermediate from which the alcoholate leaving group would depart prior to the release of both the product alcohol and inorganic phosphate from the active-site. An alternative two-step, ping-pong mechanism whereby His125 acts as the nucleophile in displacing the leaving group before being displaced from the phosphorus itself by a nucleophilic water molecule, seems to be less likely because purple acid phosphatase displays inversion of stereochemistry at the phosphorus atom.

Although each of the three histidine residues in the active-site of purple acid phosphatase (His202, His295 and His296) forms hydrogen bonds with acidic residues such that any one could act as the proton donor in the binding step, only His202 is conserved across the Ser/Thr phosphatase family. In PP1, His296 is replaced by an Arg residue and His295 is replaced by a water molecule.

The crystal structure of PP1 reveals the presence of two binding pockets that are responsible for substrate selectivity, a hydrophobic pocket, binding residues C' to the phosphorylation site, and an acidic pocket binding residues N' to the phosphorylation site, neither of which are contained within the regions of homology with other phosphatases. The hydrophobic pocket shows a prefer-

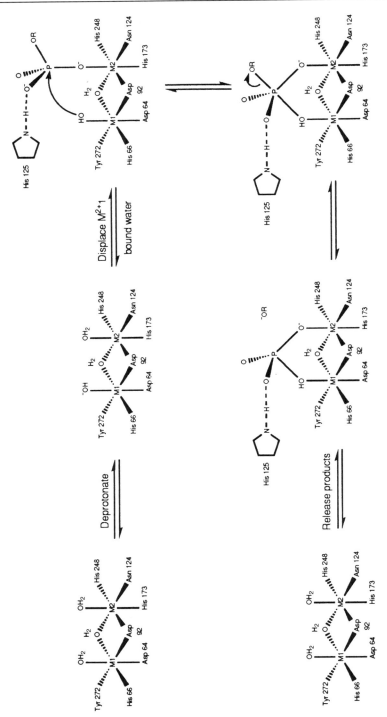

Scheme 13

ence for short-chain hydrophobic amino acids and is occupied by the Adda chain of microcystin. This Adda sidechain contains a conjugated diene moiety and is rigid, hydrophobic and extended [55] and undoubtedly contributes most of the binding energy for the inhibitor. Unlike the substrates, microcystin does not interact directly with the metal ions of the active-site, rather it is able to form a strong hydrogen bond with the water molecules in the first hydration layer of the metal ions. Thus there is no energy penalty associated with desolvating either the metal ions or the inhibitor on binding, which contributes to the very low values of K_i for these compounds.

3.4.3
Histidine Protein Phosphatases

In addition to Tyr, Ser and Thr phosphorylation, His, and to a lesser extent, Lys, phosphoproteins have been identified [64–67]. These N-phosphorylated species have escaped attention until recently because they are unstable in acid and are readily hydrolysed under the acidic conditions normally used to isolate the phosphoproteins [68]. Protein His-phosphorylation has been shown to be involved in the modulations of the mitogen-activated protein (MAP) kinase cascade, and a specific phosphatase has been implicated in the regulation cycle [65]. This phosphatase activity shows an absolute requirement for divalent metal ions and co-fractionates with the PP1/PP2A families. However, the phosphatase displays no activity towards substrates for PTPases or PP2B. It has now been demonstrated that most, if not all of the PP1/PP2A Ser/Thr protein phosphatases possess His-protein phosphatase activity, and it is unlikely that there is a separate class of enzymes that solely hydrolyse these substrates [66, 69].

3.5
5'-Nucleotidase

5'-Nucleotidase catalyses the dephosphorylation of a range of nucleoside 5'-phosphates where the preference for given substrates varies with the source of the enzyme. 5'-Nucleotidase activity was first discovered in the venom of snakes and has since been isolated from a wide variety of species, including bacteria, plants and mammals [70]. 5'-Nucleotidase activities can be split in two classes based on their substrate-binding affinities. Low-K_m nucleotidases bind 5'-IMP, 5'-AMP and 5'-GMP at micromolar concentrations, while high-K_m nucleotidases bind the same substrates at millimolar concentrations. Both classes of 5'-nucleotidase are homotetramers. The low-K_m forms possess subunit masses of ~40,000 Da, while those of the high-K_m forms range from 42–69,000 Da, depending on the source. Both types require divalent cations for activity and show a preference for Mg^{2+}. However, Ca^{2+}, Mn^{2+}, Co^{2+}, Zn^{2+} and Ni^{2+} divalent cations support catalysis with reduced activity. Both types of 5'-nucleotidase are activated by ATP, but the two forms differ in their pH optima. The high-K_m forms are maximally active at pH 6.5 while the low-K_m forms show pH optima at 7.5–9.0, depending on the structure of the 5'-nucleotide substrate. Although very similar in many respects, the two classes of 5'-nucleotidase are thought to serve different physiological roles. By virtue of its high substrate binding affinity, the low-K_m enzyme is believed to be largely responsible for the dephosphorylation of 5'-AMP. The principle substrate target for the high-K_m enzymes is believed to be 5'-IMP.

At the present time, little is known about the mechanisms of the hydrolysis reactions catalysed by these enzymes. Furthermore, no structural information has yet been reported for either class of enzyme. Nevertheless, the 5'-nucleotidases show features most closely related to those of fructose 1,6-bisphosphatase, inositol monophosphatase and the Ser/Thr protein phosphatases. For example, the enzyme from *Crotalus atrox* venom does not catalyse transphosphorylation, nor does it catalyse the exchange of ^{18}O-label from water into P_i. Thus, its properties are consistent with those expected for a ternary complex mechanism in which water directly attacks the electrophilic P atom. In keeping with the stereochemical course of all proven single step enzyme catalysed phosphoryl group transfers reported to date, the *Crotalus atrox* venom enzyme catalyses hydrolysis with inversion of configuration at the phosphorus atom [71].

Although the 5'-nucleotidases show no significant overall homology with any of the other known phosphatases, both human [72] and *Vibrio parahaemolyticus* 5'-nucleotidase [73] possess the catalytic motif characteristic of the purple acid phosphatase and Ser/Thr protein phosphatases, compared in Table 1 [56, 74–77]. There is, however, no residue corresponding to the Asp of the His-Asp pair either of the Ser/Thr protein phosphatases or of the purple acid phosphatases. In the absence of any structural data, it is not possible to say whether this omission is of any significance, as an equivalent Asp may lie elsewhere in the sequence. It is probable, therefore, that 5'-nucleotidases will show active-site structural features that are similar to those of the Ser/Thr phosphatases and the PAPs.

3.6
Inositol Monophosphatase

Inositol monophosphatase catalyses the hydrolysis of both enantiomers of myo-inositol 1- and 4-phosphate and a range of nucleoside 2'-phosphates. The enzyme has attracted considerable interest in recent years because it is believed to be an important target for lithium (cation) therapy in the treatment of manic depression [78]. Indeed, the enzyme is inhibited by Li^+ in the millimolar concentration range used in therapy and is even more sensitive to Li^+ at the physiological concentrations of the product (P_i) present in the brain (2–5 mM).

The enzyme has been purified to homogeneity from a number of mammalian sources, including rat, bovine and human brain. The proteins are very similar in size, all homodimers of about 28,000 Dalton subunits, and the bovine and human brain enzymes show only minor differences in the amino acid sequence. These enzymes have been the most intensively studied.

Inositol monophosphatase shows an absolute requirement for a divalent metal ion and activity is supported by Mg^{2+}, Mn^{2+} and Zn^{2+} ions [79]. Ca^{2+}, Gd^{3+} and a number of other divalent and trivalent metal ions are competitive inhibitors of Mg^{2+} [80, 81]. Li^+, on the other hand, inhibits uncompetitively with respect to substrate at low concentrations (double reciprocal plots of initial rate, v, versus substrate concentration, [S], give a parallel line pattern for different concentrations of Li$^+$) and noncompetitively with respect to Mg^{2+} (double reciprocal plots of v, versus [Mg^{2+}] converge on the abscissa {the negative portion of the x-axis}) [82]. These early results indicated that Li$^+$ did not simply compete for the Mg^{2+}-binding site but prevented the release of P$_i$ from the enzyme or retarded the breakdown of a possible E-P intermediate [83]. Further work demonstrated that there was a burst-phase release of inositol [78] and defined the order of product dissociation from the enzyme as inositol first and phosphate last [79]. It was also demonstrated that Mg^{2+} binds to the enzyme after the substrate and dissociates from the enzyme before P$_i$ is released [79]. However, all attempts to identify a phosphorylated enzyme species failed. Good evidence for the operation of a ternary complex mechanism which involved the direct displacement of the alcohol by a nucleophilic water molecule was obtained when it was shown that the enzyme-catalysed exchange of ^{18}O-label with P$_i$ depended absolutely on the presence of inositol [84]. Further evidence against the operation of a phosphorylated enzyme mechanism was obtained when it was shown that inositol phosphorothioates were processed by the enzyme at only slightly lower rates than the natural substrates [85]. Phosphatases that operate via E-P intermediates, including alkaline phosphatase, process phosphorothioate substrates extremely slowly. Thus, it was emerging that inositol monophosphatase operated by a very different mechanism to those of the well-studied alkaline and acid phosphatases described above.

Elegant hydroxy group deletion studies defined many of the important binding interactions of the substrate with the enzyme and demonstrated that the 3-OH and 5-OH groups of the inositol ring were not important for either binding or catalysis. Significantly, it was established that the 2-OH and the 4-OH groups and 1-O atom of the inositol skeleton were important for binding to the enzyme, while the 6-OH group was in some way involved in catalysis (Fig. 7) [86]. The publication of an X-ray crystal structure for a Gd^{3+} sulphate form of the protein allowed many of the findings from the substrate hydroxy group deletion studies to be rationalised by modelling [87]. The structure indicated clearly that the sulphate O atoms could chelate to the metal ion which was quite deeply

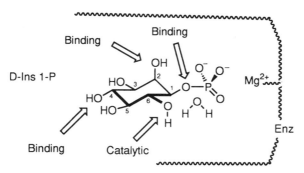

Fig. 7. A schematic representation of the enzyme-substrate interactions for Inositol Monophosphatase

buried in the enzyme. In these studies it was reasonably expected that the Gd^{3+} ion would bind in place of the Mg^{2+} ion and that the sulphate anion would serve as a surrogate for enzyme-bound P_i. However, many kinetic properties of the enzyme remained unaccounted for, including the different Mg^{2+} binding orders (timings) predicted by the crystal structure (Mg^{2+} should bind before the substrate) and observed from the kinetic studies (Mg^{2+} should bind after the substrate). Since many laboratories had recognised that the activity of the enzyme did not show simple saturation kinetics with increasing Mg^{2+} concentration, it appeared that there might be two Mg^{2+} binding sites [79].

Attempts to determine the substrate-enzyme binding interactions and identify the position of the additional Mg^{2+} binding site were pursued in two laboratories using either chemical, computational and kinetic approaches or X-ray crystallographic methods. Replacing the Gd^{3+} ion, found in the crystal structure [87], with Mg^{2+}, and sulphate with the phosphate of the substrate, allowed the reconciliation of the roles of the 2-, 4- and 6-OH groups with the structural data [89, 90]. The 4-OH group forms a hydrogen bond with the sidechain of Glu213 while the 2-OH forms a hydrogen bond with the sidechain of Asp93. A second magnesium ion binding site, $Mg^{2+}2$, is formed by Asp90, Asp93 and Asp220, and a metal ion bound there can be coordinated by both the phosphate ester-bridging oxygen atom and one of the non-bridging phosphate O atoms (Fig. 8a). In this conformation, the 6-OH of inositol does not interact directly with the metal ion in the $Mg^{2+}2$ site, rather it is able to form a hydrogen bond with a water molecule that completes the primary coordination sphere of $Mg^{2+}2$. The chemical approach concerned the enzyme's ability to process nucleoside 2′-phosphates and 2′-thiophosphates which was probed using the Mg^{2+} and the Mn^{2+} forms of the enzyme [88]. The results of these studies and the finding that adenosine was unable to mediate the exchange of ^{18}O-label from water with P_i (unlike inositol, see above) was used to determine the active conformation for adenosine 2′-phosphate at the active-site of the enzyme. It was initially proposed by Cole and Gani [88] that the 4′-hydroxymethyl group and the adenine moiety occupied unfavourable axial positions in the ribofuranosyl ring. They reasoned that only chelation of the ether O atom and the (leaving) 2′-O atom of the ribofuranosyl moiety by a metal ion to form a five-membered metallocycle would sufficiently stabilise the conformation. However, calculation of the conformation of 2′-AMP bound in the active-site of inositol monophosphatase showed that while the phosphate group of 2′-AMP is bound in a similar position and orientation to that adopted by the phosphate group of D-inositol 1-phosphate, the ether O atom of the ribose ring cannot interact directly with the metal ion bound in the $Mg^{2+}2$ site [89]. An alternate arrangement, whereby the ribofuranosyl ring O atom interacted indirectly with the $Mg^{2+}2$-bound water molecule via a second water molecule (see Fig. 8b) was also consistent with the absence of adenosine-mediated ^{18}O exchange. In this case, the internal conformational energy of the substrates is low, but the requirement for two bound water molecules reduces the overall entropy of the system. In addition, the 5′-OH group of 2′-AMP forms a hydrogen bond with Glu213. Thus, a role for the second Mg^{2+} became evident. Comparison with the analogous conformations for inositol phosphates indicated that the 1-O atom (the nucleofuge)

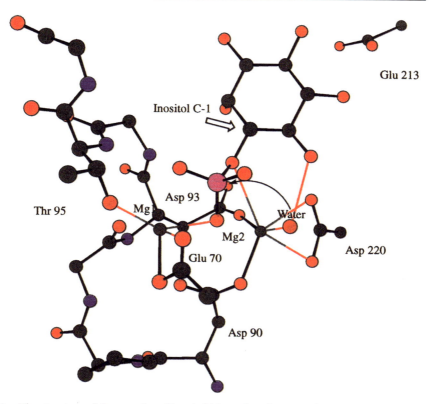

Fig. 8a. The structure of the complex of Inositol Monophosphatase and D-myo-Inositol phosphate with the attacking nucleophile bound to $Mg^{2+}2$

and the nucleophile should chelate to the second Mg^{2+} ion. The catalytically essential 6-OH group serves both to aid in the binding of the nucleophile and to present it in the correct orientation for attack at the P atom.

The finding that two Mg^{2+} ions were required and their identified actual positions in the active complex explained several unanswered questions, including why a catalytically important 6-OH site existed in the inositol skeleton, why there were discrepancies in the perceived binding orders for Mg^{2+} and why Li^+ inhibition changed from simple uncompetitive at low $[Li^+]$ to mixed at high $[Li^+]$. Subsequent studies using substrates and inhibitor analogues provided further evidence for the operation of a two metal site mechanism (see Fig. 8) [89, 90].

Independently, X-ray crystallographic studies of protein-substrate complexes using inhibitory metals to prevent reaction provided an almost identical picture of the active-site interactions (Fig. 9) [91]. A Gd^{3+}-inhibited complex [92] containing the natural substrate, D-inositol 1-phosphate, showed the same binding interactions between the enzyme and substrate found in the computational study [89]. Only one metal ion was found in this structure, occupying the

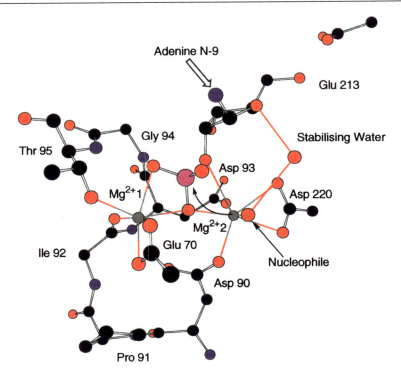

Fig. 8b. The structure of the complex of Inositol Monophosphatase and 2′-AMP, including both the nucleophilic water molecule and the second, catalytically important water molecule

$M^{2+}1$ site with an occupancy of 35%. In this case however, Li^+ was present in the crystallisation solvent and may have occupied the second site. Further structures containing Mn^{2+} or Mn^{2+} and phosphate [93] showed a second metal ion bound in the predicted $M^{2+}2$ site. Surprisingly, the Mn^{2+} structure showed a third metal ion binding site, though this site is occupied by a water molecule in the presence of phosphate. Indeed, only one difference in the proposed mechanisms existed. This concerned which of the two metal ion sites provided activation, through chelation, for the nucleophilic water molecule. The crystal structures of inositol monophosphatase [87, 92, 93] showed a water molecule coordinated to $Mg2+1$ and positioned close to the P atom ready for inline nucleophilic attack. This water molecule also formed hydrogen bonds with the sidechains of Thr95 and Glu70.

On the basis of the rapid rates of ^{18}O-exchange from water into P_i in the presence of inositol and the established orders for metal ions binding to the enzyme, Cole and Gani argued that the second Mg^{2+} ion should activate the water molecule [88]. This was because both $Mg^{2+}2$ and its associated water molecules could readily dissociate from the enzyme and exchange with the solvent and rebind without the substrate dissociating. The X-ray study placed the nucleophilic water molecule on the first Mg^{2+}. The two different sites for the

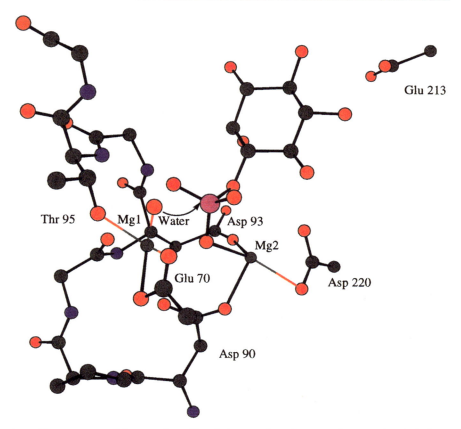

Fig. 9. The structure of the complex of inositol monophosphatase and D-myo-inositol phosphate with the attacking nucleophile bound to $Mg^{2+}1$

nucleophile would give different stereochemical courses for phosphoryl transfer. A location on $Mg^{2+}1$ would give inversion of configuration through an in-line displacement (Scheme 5 A, B or C) whereas attack by a water molecule chelated to $Mg^{2+}2$ would give retention via an adjacent association and pseudorotation (Scheme 5D). Computational determination of the reaction pathways for models of these alternative mechanisms have shown that it is not possible to differentiate between them on energetic grounds [63]. Under certain circumstances, the activation energy for the non-inline process could be lower than that for the corresponding inline process. The stereochemical course of the reaction has not yet been determined, but work has been carried out to design molecules capable of inhibiting one mechanism but not the other.

Specifically, the 6-hydroxyl of the substrate, which lies close to $Mg^{2+}2$, has been replaced with a longer pendant arm hydroxyl group to give compound (**4**) that is able to coordinate $Mg^{2+}2$ and displace any water molecule bound there (Fig. 10) [89, 94, 95]. Should $Mg^{2+}2$ activate the nucleophile, this pendant arm

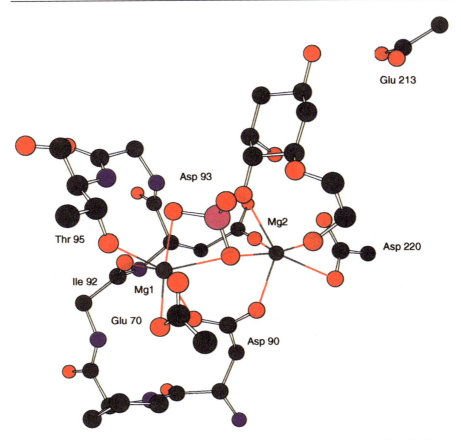

Fig. 10. The structure of the complex of inositol monophosphatase and the O^6-(2'-hydroxy-ethyl)-inositol 1-phosphate analogue. The analogue is bound with the 2'-OH group coordinating $Mg^{2+}2$ and displacing the nucleophilic water molecule

compound would function as an inhibitor of inositol monophosphatase or, even as a proposed nucleophilic water has been replaced by a comparable nucleophilic hydroxyl, act as a substrate for a transphosphorylation reaction whereby the phosphate group is transferred from the 1 position to the 2'-O atom of the pendant arm on the same molecule, to give the primary alkyl phosphate (5). It turns out that compound (4) acts as a potent competitive inhibitor of inositol monophosphates, as does the phosphorylated pendant arm isomer (5), though neither can function as a substrate for an intramolecular transphosphorylation [95]. Interestingly, the conformationally restricted cyclic phosphate (6), which

O^6-(2'-hydroxyethyl)-Inositol 1-Phosphate Analogue
(**4**)

O^6-pendant arm transphosphorylation product
(**5**)

O^6-cyclic Inositol 1-Phosphate Analogue
(**6**)

is a transition state analogue for the transphosphorylation process, is the most potent monoanionic phosphate inhibitor known for inositol monophosphatase. These results are entirely consistent with the nucleophile originating from $Mg^{2+}2$, rather than $Mg^{2+}1$. If the latter metal ion activated the nucleophile, the 2'-hydroxyethyl phosphate (**4**) and other O^6-alkyl substituted phosphates would be expected to act as substrates.

3.7
D-Fructose 1,6-Bisphosphate 1-Phosphatase

Fructose 1,6-bisphosphatase hydrolyses the 1-phosphate ester group of D-fructose 1,6-bisphosphate to give fructose 6-phosphate and P_i. The enzyme is a homotetramer consisting of 35,000-Dalton subunits. Kinetic studies show that only the α-anomer of the substrate is hydrolysed, but a slow, non-enzymic interconversion of the α- and β-anomers ensures that the complete hydrolysis of a mixture of the anomers occurs with time [96, 97]. The enzyme is inhibited allosterically by 5′-AMP and is activated by divalent metal cations such as Mg^{2+}, Mn^{2+} and Zn^{2+}. As is the case with inositol monophosphatase, these ions are inhibitory at high concentrations. Kinetic studies of metal ion binding have shown that between one and three Zn^{2+} ions per monomer are able to bind. For Mn^{2+}, in the absence of substrate, only a single ion is able to bind, but a second ion can bind in the presence of the substrate or a substrate analogue. There is still some uncertainty regarding the binding of Mg^{2+} where there appears to be two binding sites at pH 7.2 and 9.1 in the absence of EDTA, but only a single site in the presence of EDTA. Current opinion holds that two metal ions are required for catalysis [98] and that the E. Mg^{2+}.S complex possesses a very low affinity for a second Mg^{2+} ion. This property might explain why crystal structures have been determined showing two Zn^{2+} ions or two Mn^{2+} ions at the active-site, but only singly occupancy for Mg^{2+} [98, 99].

As is the case for inositol monophosphatase, no phosphoenzyme intermediate could be detected. Fructose bisphosphatase is able to catalyse ^{18}O exchange from ^{18}O-labelled P_i into the solvent in the absence of the product, fructose 6-phosphate, although the reaction is extremely slow, 160 times slower than in the presence of fructose 6-phosphate ($t_{1/2} = 14$ min.) [100] The ^{18}O-exchange reaction requires the presence of a divalent cation, shows the same broad specificity for M^{2+} as the hydrolysis reaction and a maximum rate when the enzyme is saturated with both P_i and fructose 6-phosphate. These properties together are indicative of a direct displacement ternary complex mechanism. Benkovic and co-workers have shown that the hydrolysis reaction proceeds with inversion of configuration at the phosphorus atom [101] which further refines the geometry of displacement by water or hydroxide as of the in-line type (see mechanisms B or C of Scheme 5).

Several crystal structures of fructose bisphosphatase have been determined [98, 99] with a variety of metal cations and substrate analogues bound at the active-site. The absence of any likely protein-based nucleophile in the vicinity of the 1-phosphate (Fig. 11) supports the view that this enzyme, like inositol monophosphatase, conducts hydrolysis by means of a direct displacement.

Despite the clear similarities in mechanism between fructose bisphosphatase and inositol monophosphates described above, the two enzymes show very little sequence and structural homology. Their different substrate specificities account for much of this, but a comparison of the metal binding sequences is of special interest. The crystal structure of inositol monophosphatase shows the sequence 90–96 (Asp-Pro-Ile-Asp-Gly-Thr-Thr) forming a kinked struc-

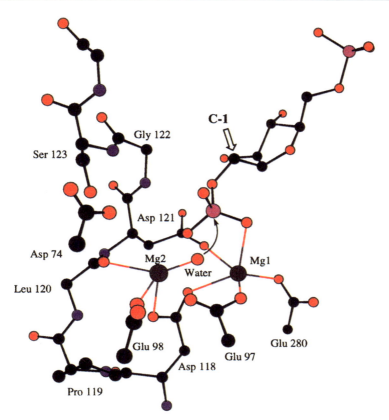

Fig. 11. The structure of the complex of fructose bisphosphatase and the competitive inhibitor, 2,5-anhydroglucitol 1,6-bisphosphate

ture, binding the metal (Gd^{3+}) with the sidechain of Asp90, the backbone carbonyl group of Ile92 and the sidechain of Thr95. The corresponding sequence in fructose bisphosphatase, 118–124 (Asp-Pro-Leu-Asp-Gly-Ser-Ser), takes up an identical conformation (see Fig. 11) and contributes to the binding at both metal ion sites. The sidechains of Asp118 and Asp121 coordinate to M^{2+} ions at the $M^{2+}1$ site (the only site shown to bind Mg^{2+}), while the sidechain of Asp118 is also involved in chelating to the M^{2+} ion at the $M^{2+}2$ site, along with the backbone carbonyl group of Leu120 and the sidechain of Ser123 (Fig. 11). Sidechains of acid residues from elsewhere in the chain complete the coordination shells at the two sites, but do not show any sequence homology with equivalent residues from inositol monophosphatase.

Note that the $M^{2+}2$ site of fructose bisphosphatase corresponds positionally to the first metal ion binding site of inositol monophosphatase. The $M^{2+}1$ site is similar to the second Mg^{2+}-binding site of inositol monophosphatase, suggesting that these two enzymes are mechanistically similar. However, in contrast to the situation for inositol monophosphatase, the O atom which dissociates from the substrate phosphomonoester does not appear to be chelated by either of the metal ions. Thus, for the reverse reaction, the pK_a value for the alcohol nucleophile is not modified by chelation to a metal ion. This apparent diversity, if borne out, represents a major difference in the mechanism of the two enzymes and might explain why fructose 1,6-bisphosphatase can catalyse the slow exchange of ^{18}O-label from ^{18}O-water into inorganic phosphate in the absence of the co-product alcohol. If the O atom which dissociates is chelated in the active complex for fructose 1,6-bisphosphatase, it would be through an interaction with the ion in the $M^{2+}1$ site rather than with the second binding ion, as is the case for inositol monophosphatase.

4
Overview

It is evident from this cursory examination of the mode of action of the mono-phosphate ester hydrolase enzymes that nature has evolved many different mechanisms for cleaving P-O bonds. Interestingly, the less specific enzymes, including alkaline and acid phosphatase, operate via phosphorylated enzyme intermediates. At a chemical level, this may reflect the fact that there is very little to bind to in catalysing a non-specific hydrolysis, just the phosphate group. Therefore, in order to standardise the catalytic problem, the enzyme might be expected to first transphosphorylate internally to release the alcohol product and then tackle the hydrolytic step on a species which is identical, regardless of the starting substrate. Specific enzymes can, of course, bind as much of the substrate as is required for stabilising the transition state for the direct hydrolysis by water, or hydroxide, in a single step.

While the type and geometrical arrangement of the metal ions used in phosphohydrolase catalysis varies widely (see Fig. 12), in each case their roles are very similar; to position and activate the nucleophile, to position and enhance the elecrophilicity of the P atom and to provide Lewis acid catalysis for the leaving alkoxide (where necessary). The reported stereochemical courses of all of the systems examined above show that single phosphoryl transfer steps occur with inversion and, where retention occurs, there are an even number of transfer steps. Whether inositol monophosphatase, which acts via a direct displacement, will follow this trend remains to be seen.

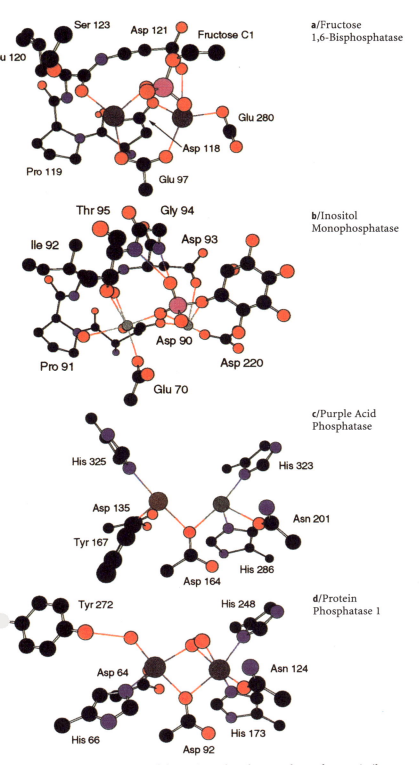

a/Fructose 1,6-Bisphosphatase

Ser 123 · Asp 121 · Fructose C1 · Leu 120 · Glu 280 · Asp 118 · Pro 119 · Glu 97

b/Inositol Monophosphatase

Thr 95 · Gly 94 · Ile 92 · Asp 93 · Asp 90 · Pro 91 · Asp 220 · Glu 70

c/Purple Acid Phosphatase

His 325 · His 323 · Asp 135 · Asn 201 · Tyr 167 · His 286 · Asp 164

d/Protein Phosphatase 1

Tyr 272 · His 248 · Asp 64 · Asn 124 · His 66 · His 173 · Asp 92

Fig. 12. The active site structures of the various phosphatases shown from a similar perspective, relative to the metal ions and/or phosphate group (or surrogate)

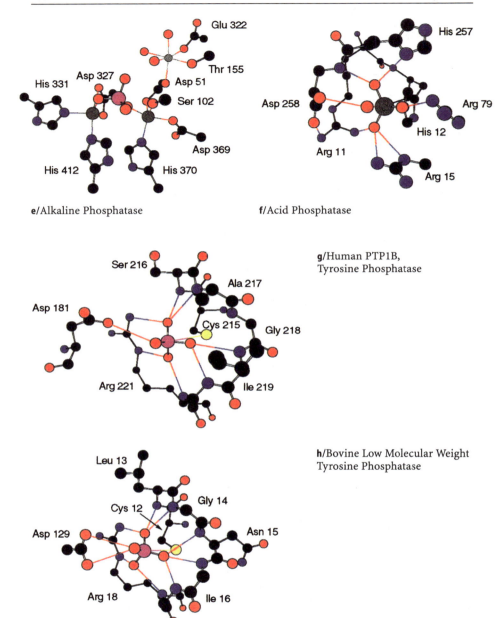

e/Alkaline Phosphatase

f/Acid Phosphatase

g/Human PTP1B,
Tyrosine Phosphatase

h/Bovine Low Molecular Weight
Tyrosine Phosphatase

Fig. 12 (continue)

5
References

1. Krebs EG, Fischer EH (1956) Biochim Biophys Acta 20:150
2. Bruice TC, Benkovic SJ (1966) in Bioorganic mechanisms Vol. 2, W. A. Benjamin, New York, pp 1–109
3. Perrin DD, Dempsey B (1974) in Buffers of pH and metal ion control, Chapman and Hall, London, p 104
4. Dougherty TM, Clelland WW (1985) Biochemistry 24:5870
5. Mildvan AS, Cohn M (1965) J Biol Chem 240:238
6. Herschlag D, Jencks WP (1990) Biochemistry 29:5172
7. Herschlag D, Jencks WP (1986) J Am Chem Soc 108:7938
8. Herschlag D, Jencks WP (1989) J Am Chem Soc 111:7579
9. Herschlag D, Jencks WP (1989) J Am Chem Soc 111:7587
10. Herschlag D, Jencks WP (1990) J Am Chem Soc 112:1942
11. Herschlag D, Jencks WP (1990) J Am Chem Soc 112:1951
12. Knowles JR (1980) Annu Rev Biochem 49:877
13. Floss HG, Tsai MD, Woodard RW (1984) Top Stereochem 15:253
14. Frey PA (1989) Adv Enzymol Relat Areas Mol Biol 62:119
15. Buchwald SL, Pliura DH, Knowles JR (1984) J Am Chem Soc 106:4916
16. Anderson RA, Bosron WF, Kennedy FS, Vallee BL (1975) Proc Natl Acad Sci 72:2989
17. Jones SR, Kindman LA, Knowles JR (1978) Nature 257:564
18. Han R, Coleman JE (1995) Biochemistry 34:4238
19. Dayan J, Wilson IB (1964) Biochim Biophys Acta 81:620
20. Wilson IB, Dayan J, Cyr K (1964) J Biol Chem 239:4182
21. Fernley HN, Walker PG (1966) Nature 212:1435
22. Aldridge WN, Barman TE, Gutfreund H (1964) Biochem J 92:23C
23. Schwartz JH, Crestfield AM, Lipmann F (1963) Proc Natl Acad Sci 49:722
24. Kim EE, Wyckoff HW, (1991) J Mol Biol 218:449
25. Sowadski JM, Handschumacher MD, Murthy H, Foster BA, Wyckoff HW (1985) J Mol Biol 186:417
26. Van Etten RL (1982) Ann New York Acad Sci 390:27
27. Ostanin K, Harms EH, Stevis PE, Kuciel R, Zhou MM, Van Etten RL (1992) J Biol Chem 267:22830
28. Buchwald SL, Saini MS, Knowles JR, Van Etten RL (1984) J Biol Chem 259:2208
29. Van Etten RL, Risley JM (1978) Proc Natl Acad Sci 75:4784
30. Fothergill-Gilmore LA, Watson HC (1989) Adv Enzymol Relat Areas Mol Biol 62:227
31. Lindqvist Y, Schneider G, Vihko P (1994) Eur J Biochem 221:139
32. Lindqvist Y, Schneider G, Vihko P (1993) J Biol Chem 268:20744
33. Pyrz JW, Sage JT, Brunner PG, Que L (1986) J Biol Chem 261:11015
34. Strater N, Klabunde T, Tucker P, Witzel H, Krebs B (1995) Science 268:1489
35. Beck JL, McConachie LA, Summors AC, Arnold WN, De Jersey J, Zerner B (1986) Biochim Biophys Acta 869:61
36. Vincent JB, Crowder MW, Averill BA (1991) J Biol Chem 266:17737
37. Mueller EG, Crowder MW, Averill BA, Knowles JR (1993) J Am Chem Soc 115:2974
38. Doi K, McCracken J, Peisach J, Aisen P (1988) J Biol Chem 263:5757
39. Klabunde T, Strater N, Frohlich R, Witzel H, Krebs B (1996) J Mol Biol 259:737
40. Charbonneau H, Tonks NK (1992) Ann Rev Cell Biol 8:469
41. Fischer EH, Charbonneau H, Tonks NK (1991) Science 253:401
42. Walton KM, Dixon JE (1993) Ann Rev Biochem 62:101
43. Saini MS, Buchwald SL, Van Etten RL, Knowles JR (1981) J Biol Chem 256:10453
44. Denu JM, Lohse DL, Vijayalakshmi J, Saper MA, Dixon JE (1996) Proc Natl Acad Sci 93:2493
45. Cirri P, Chiarugi P, Camici G, Manou G, Raugei G, Cappugi G, Ramponi G (1993) FEBS Lett 214:647

46. Zhang M, Van Etten RL, Stauffacher CV (1994) Biochemistry 33:11097
47. Logan TM, Zhou MM, Nettesheim DG, Meadows RP, Van Etten RL, Fesik SW (1994) Biochemistry 33:11087
48. Su XD, Taddei N, Stefani M, Ramponi G, Nordlund P (1994) Nature 370:575
49. Taddei N, Chiarugi P, Cirri P, Fiaschi T, Stefani M, Camici G, Raugei G, Ramponi G (1994) FEBS Lett 350:328
50. Wu L, Zhang AY (1996) Biochemistry 35:5426
51. Barford D, Flint AJ, Tonks NK (1994) Science 263:1397
52. Jia ZC, Barford D, Flint AJ, Tonks NK (1995) Science 268:1754
53. Cohen P (1989) Ann Rev Biochem 58:453
54. Pilgrim DB, McGregor A, Jaekle P, Johnson T, Hansen D, EMBL/Genbank/DDBJ databank, ref FEM2_CAEEL.
55. Goldberg J, Huang HB, Kwon YG, Greengard P, Nairn AC, Kuriyan J (1995) Nature 376:745
56. Zhuo S, Clemens JC, Stone RL, Dixon JE (1994) J Biol Chem 269:26234
57. Griffith JP, Kim JL, Kim EE, Sintchak MD, Thomson JA, Fitzgibbon MJ, Flemming MA, Caron PR, Hsiao K, Navia MA (1995) Cell 82:507
58. Mehrotra AP, Gani D (1996) Tetrahedron Lett 37:6915
59. Martin BL, Graves DJ (1994) Biochim Biophys Acta 1206:136
60. Martin B, Pallen CJ, Wang JH, Graves DJ (1985) J Biol Chem 260:14932
61. Martin BL, Graves DJ (1986) J Biol Chem 261:14545
62. Vincent JB, Crowder MW, Averill BA (1992) Biochemistry 31:3033
63. Wilkie J, Gani D (1996) J Chem Soc Perkin Trans 2 783
64. Rajagopal P, Waygood EB, Klevit RE (1994) Biochemistry 33:15271
65. Maeda T, Wurgler-Murphy SM, Salto H (1994) Nature 369:242
66. Motojima K, Goto S (1994) J Biol Chem 269:9030
67. Wong C, Faiola B, Wu W, Kennelly PJ (1993) Biochem J 296:293
68. Chen CC, Bruegger BB, Kern CW, Lin YC, Halpern RM, Smith RA (1977) Biochemistry 16:4852
69. Matthews HR, MacKintosh C (1995) FEBS Lett 364:51
70. Itoh R (1993) Comp Biochem Physiol 105B:13
71. Tsai MD, Chang TT (1980) J Am Chem Soc 102:5416
72. Misumi Y, Ogata S, Ohkubo K, Hirose S, Ikehara Y (1990) Eur J Biochem 191:563
73. Tamao Y, Noguchi K, Sakai-Tomita Y, Hama H, Shimamoto T, Kanazawa H, Tsuda M, Tsuchiya T (1991) J Biochem 109:24
74. Menzel D, Vugrek P, Frank S, Elsner-Menzel C (1995) Eur J Cell Biol 67:179
75. Arino J, Perez-Callejon E, Cunillera N, Camps M, Posas F, Ferrer A (1993) Plant Mol Biol 21:475
76. Guerini D, Montell C, Klee CB (1992) J Biochem 267:22542
77. Klabunde T, Stahl B, Suerbaum H, Hahner S, Karas M, Hillenkamp F, Krebs B (1994) Eur J Biochem 226:369
78. Gani D, Downes CP, Batty I, Bramham J (1993) Biochim Biophys Acta 1177:253
79. Leech AP, Baker GR, Shute JK, Cohen MA, Gani D (1993) Eur J Biochem 212:693
80. Hallcher LM, Sherman WR (1980) J Biol Chem 255:10896
81. Takimoto K, Okada M, Matsuda Y, Nakaga H (1985) J Biochem 98:363
82. Ganzhorn AJ, Chanal MC (1990) Biochemistry 29:6065
83. Shute JK, Baker R, Billington DC, Gani D (1988) J Chem Soc Chem Comm 626
84. Baker GR, Gani D (1991) Bio Med Chem Lett 1:193
85. Baker GR, Billington DC, Gani D (1991) Bio Med Chem Lett 1:17
86. Baker R, Carrick C, Leeson PD, Lennon IC, Liverton N (1991) J Chem Soc Chem Comm 298
87. Bone R, Springer JP, Atack JR (1992) Proc Natl Acad Sci. 89:10031
88. Cole AG, Gani D (1994) J Chem Soc Chem Comm 1139
89. Wilkie J, Cole AG, Gani D (1995) J Chem Soc Perkin Trans 1 2709
90. Cole AG, Wilkie J, Gani D (1995) J Chem Soc Perkin Trans 1 2695

91. Pollack SJ, Atack JR, Knowles MR, McAllister G, Ragan CI, Baker R, Fletcher SR, Iversen LL, Broughton HB (1994) Proc Natl Acad Sci 91:5766
92. Bone R, Frank L, Springer JP, Pollack SJ, Osborne SA, Atack JR, Knowles MR, McAllister G,.Ragan CI, Broughton HB, Baker R, Fletcher SR (1994) Biochemistry 33:9460
93. Bone R, Frank L, Springer JP, .Atack JR (1994) Biochemistry 33:9468
94. Schultz J, Wilkie J, Lightfoot P, Rutherford T, Gani D (1995) J Chem Soc Chem Comm 2353
95. Schultz J, Gani D (1997) J Chem Soc Perkin Trans 1 657
96. Benkovic PA, Bullard WP, de Maine MM, Fishbein R, Schray KJ, Steffens JJ, Benkovic SJ (1974) J Biol Chem 249:930
97. Frey WA, Fishbein R, de Maine MM, Benkovic SJ (1977) Biochemistry 16:2479
98. Zhang Y, Liang JY, Huang S, Ke H, Lipscomb WN (1993) Biochemistry 32:1844
99. Ke H, Zhang Y, Lipscomb WN (1990) Proc Natl Acad Sci 87:5243
100. Sharp TR, Benkovic SJ (1979) Biochemistry 18:2910
101. Domanico PL, Rahil JF, Benkovic SJ (1985) Biochemistry 24:1623

The Dimetal Center in Purple Acid Phosphatases

Thomas Klabunde[1] and Bernt Krebs[2]*

[1] Department of Biochemistry and Biophysics, Texas A&M University, College Station, TX 77483–2128, USA, *E-mail: Klabund@monoc.tamu.edu*
[2] Institut für Anorganische Chemie, Universität Münster, Wilhelm-Klemm Str. 8, 48149 Münster, Germany, *E-mail: Krebs@nwz.uni-muenster.de*

Purple acid phosphatases (PAPs) contain a dinuclear metal center in their active site and hydrolyze phosphoric acid esters at low pH. Characteristic of this group of acid phosphatases is their resistance to inhibition by tartrate and their purple color, due to the presence of a tyrosine residue ligated to a ferric iron. The mammalian enzymes all contain a mixed-valent di-iron unit in their catalytic active form, first identified in the bovine spleen and porcine uterus enzymes, while a heterodinuclear Fe(III)Zn(II) unit has been characterized for the most studied plant enzyme from kidney bean. The enzymes from porcine uterus and bovine spleen can be converted into active FeZn forms and the plant enzyme can be transformed into an active FeFe form. In recent years the dimetal center of PAPs has been studied using numerous spectroscopic methods such as Mössbauer spectroscopy, EPR, NMR, EXAFS, magnetic, electrochemical and resonance Raman studies characterizing most of the metal coordinating residues, the metal-metal separation and providing evidence of the similarity between enzymes from different sources. Analysis of the products of hydrolysis of a substrate containing a chiral phosphorus by ^{31}P NMR, stopped-flow measurements and kinetic studies all support a reaction path involving nucleophilic attack of a Fe(III)-bound hydroxide ligand on the phosphate ester. The recently solved crystal structure of the plant enzyme provides the structural basis for the understanding of the two-metal ion mechanism of this class of enzymes.

Keywords: Purple acid phosphatase, tartrate-resistant acid phosphatase, metalloenzyme, di-iron protein.

List of Abbreviations 178

1   Introduction: About PAPs and TRAPs 178

2   Molecular Properties 179

2.1 Molecular Weight and Amino Acid Sequence: Do PAPs
     from Different Sources Form a Single Entity ? 179
2.2 Overall Structure and Protein Folding of the Plant PAP 181

3   The Dimetal Site 183

3.1   Spectroscopic Characterization 183
3.1.1 UV/Vis and Resonance Raman Spectra 183
3.1.2 EPR and Magnetic Measurements 183

---

* Corresponding Author.

3.1.3   ESEEM and ENDOR Spectra 186
3.1.4   Mössbauer Spectra 186
3.1.5   EXAFS Spectra 186
3.1.6   ¹H-NMR Spectra 186
3.2   The Crystal Structure of the Fe(III)Zn(II) Unit of kbPAP 189
3.3   Interaction with Phosphate and Other Tetrahedral Oxoanions 190

4   The Proposed Reaction Path: Two-Metal Ion Catalysis by PAPs .. 193

5   The Enigma: The Physiological Role of PAP/TRAPs 195

6   References 195

List of Abbreviations

Uf   uteroferrin
afPAP   *Aspergillus ficuum* purple acid phosphatase
ncPAP   *Neurospora crassa* purple acid phosphatase
bsPAP   bovine spleen purple acid phosphatase
kbPAP   kidney bean purple acid phosphatase
sbPAP   soy bean purple acid phosphatase
spPAP   sweet potato purple acid phosphatase
PAP   purple acid phosphatase
TRAP   tartrate-resistant acid phosphatase
SQUID   superconducting quantum interference device
ENDOR   electron nuclear double resonance
ESSEM   electron spin echo envelope modulation
EXAFS   extended X-ray absorption fine structure

1
Introduction: About PAPs and TRAPs

Acid phosphatases represent a group of enzymes that hydrolyze phosphoric acid esters and show optimal catalytic activity at a low pH. A purple-colored, iron-containing subclass of acid phosphatases was first isolated from bovine spleen (bsPAP) and porcine uterine fluid (uteroferrin, Uf) [1, 2]. Because of their intense color these acid phosphatases have come to be known as purple acid phosphatases (PAPs) (for previous reviews see Refs. [3–5]). Almost at the same time medical researchers characterized an acid phosphatase from human tissues which is resistant to inhibition by tartrate [6–8]. They termed this enzyme human type 5 (based on the electrophoretic mobility), tartrate-resistant acid phosphatase (TRAP). Despite the different names given to these enzymes by researchers in both areas, they show significant sequence homology [9], and immunological studies [10] also reveal that the PAPs from animal sources and the human TRAPs are in fact identical.

In recent years PAPs have been isolated from a variety of animal, plant and bacterial sources. Besides bsPAP and Uf, mammalian PAPs have been characterized from different human tissues (spleen [6–8], placenta [9], bone [11], lung [12, 13]), bovine bone [14], rat spleen [15], bone [16], and epidermis [17]. The dimetal sites from bsPAP and Uf especially have been studied using numerous spectroscopic techniques, providing evidence for the close similarity of both active sites and yielding a well-defined model of the di-iron unit. The mammalian PAPs resemble other di-iron proteins like hemerythrin, ribonucleotide reductase, methane monooxygenase, rubrerythrin, and the stearyl acyl carrier protein Δ^9-desaturase, in that they contain an antiferromagnetically coupled di-iron unit. The 3-dimensional crystal structures of all these proteins involved in transport or activation of oxygen have been recently determined (structural features of their di-iron units are reviewed in Ref. [18]).

The class of PAPs from plants includes dimeric glycoenzymes from kidney bean seeds (kbPAP, [19, 20]), soybean seedlings (sbPAP, [21, 22]), sweet potato tubers (spPAP, [23, 24]), spinach leaves [25], rice-cultured cells [26], and *Arabidopsis thaliana* [27]. Recently the X-ray crystal structure of the best-studied FeZn plant PAP from kidney beans has been determined at a resolution of 2.65 Å [28, 29]. Interestingly, the FeZn kidney bean enzyme can be converted into an active di-iron form [30], and the enzymes from porcine uterus and bovine spleen can also be transformed into active FeZn forms [31, 32]. Less is known about the metal cofactors of PAPs present in the cells of other plants. Two different groups have reported that the enzyme from sweet potato tuber from different sources contains either manganese or iron.

Bacterial purple phosphatases with a pH optimum of about 6.0 have been isolated from *Aspergillus ficuum* (afPAP, [33]) and *Neurospora crassa* (ncPAP, [34]). The absorption maximum at 580 nm reported for the afPAP means this enzyme has a blue rather than a purple color. Unfortunately, the metal content of bacterial PAPs is as yet unknown. A purple phosphatase from *Micrococcus sodenensis* shows maximal activity in the alkaline range [35]. Too little is known about the properties of this enzyme, which is therefore not discussed in this review.

2
Molecular Properties

2.1
Molecular Weight and Amino Acid Sequence: Do PAPs from Different Sources Form a Single Entity ?

Despite the characteristic features that warrant their designation as purple acid phosphatases, these enzymes from mammalian, plant and bacterial species differ considerably in their molecular mass, subunit arrangement, and amino acid sequence. Whereas the mammalian enzymes are monomeric glycoproteins with a molecular mass of approximately 35 kDa, in contrast the molecular mass of the dimeric plant enzymes ranges from 100 to 120 kDa. The two characterized bacterial purple acid phosphatases afPAP and ncPAP represent highly glycosylated monomeric proteins with a molecular mass of 82 and 85 kDa, respectively.

However, an alignment of the plant and mammalian sequences revealed a local sequence homology restricted to five regions [36]. This study strongly sug-

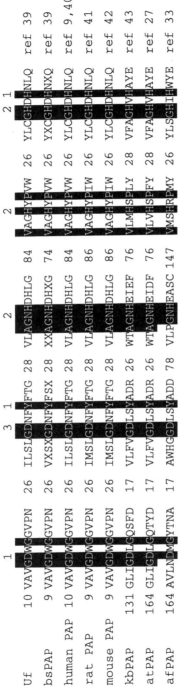

```
                              1                    3                      2                         2                 1
                                                                                                                      2
Uf          10 VAVGDWGGVPN 26 ILSLGDNFYFTG 28 VLAGNHDHLG 84 VAGHYPVW 26 YLCGHDHNLQ   ref 39
bsPAP        9 VAVGDWGGVPN 26 VXSXGDNFYFSX 28 XXAGNHDHXG 74 VAGHYPVW 26 YXCGHDHNXQ   ref 39
human PAP   10 VAVGDWGGVPN 26 ILSLGDNFYFTG 28 VLAGNHDHLG 84 VAGHYPVW 26 YLCGHDHNLQ   ref 9,40
rat PAP      9 VAVGDWGGVPN 26 IMSLGDNFYFTG 28 VLAGNHDHLG 86 VAGHYPIW 26 YLCGHDHNLQ   ref 41
mouse PAP    9 VAVGDWGGVPN 26 IMSLGDNFYFTG 28 VLAGNHDHLG 86 VAGHYPIW 26 YLCGHDHNLQ   ref 42
kbPAP      131 GLIGDLGQSFD 17 VLFVGDLSYADR 26 WTAGNHEIEF 76 VLMHSPLY 28 VFAGHVHAYE   ref 43
atPAP      164 GLIGDLGQTYD 17 VLFVGDLSYADR 26 WTAGNHEIDF 76 VLVHSPFY 28 VFAGHVHAYE   ref 27
afPAP      164 AVLNDMGYTNA 17 AWHGGDLSYADD 78 VLPGNHEASC 147 VMSHRPMY 26 YLSGHIHWYE   ref 33
```

Fig. 1. Sequence alignment of purple acid phosphatases showing the conserved metal-ligating residues. Complete amino acid sequences are available for numerous mammalian PAPs (Uteroferrin [39], bovine spleen PAP [39], human [9, 40], rat [41], and mouse PAP/TRAP [42]), revealing a sequence homology of over 90%, for the plant enzymes from *Phaseolus vulgaris* (kbPAP) and *Arabidopsis thaliana* [27] (68% identical residues) and for a single bacterial PAP from *Aspergillus ficuum* [33]. Those residues coordinated to metal ion **1** (Fe in kbPAP) and **2** (Zn in kbPAP) are marked. The numbers on the left indicate the amino acid position of the first residue. The numbers between regions show the number of amino acid residues separating these regions. Note the different size of the separating regions indicating the presence of additional domains in the plant and bacterial PAP compared to the mammalian enzymes gests that all metal-ligating residues identified in the crystal structure of the kidney bean enzyme are conserved in the mammalian PAPs (Fig. 1). Similarly, a local alignment of the sequences of the bacterial afPAP with the human type 5 TRAP reveals regions with significant sequence homology [33]. The conserved fragments also contain the metal-ligating residues of the kidney bean enzyme, suggesting that mammalian, plant, and bacterial PAPs are related and share identical metal-ligating residues (but differ in the metal content of the native enzyme form). Furthermore, VIS, EPR, proton NMR and Mössbauer spectroscopy data on bsPAP, Uf, and the Zn-exchanged di-iron kbPAP indicate a high degree of conformity in active sites of these enzymes [37, 38].

2.2
Overall Structure and Protein Folding of the Plant PAP

Determination of the 3-dimensional crystal structure of the kidney bean purple acid phosphatase revealed the molecular architecture of this 111-kDa plant enzyme [28, 29]. The homodimeric enzyme, with a disulfide bridge connecting the two subunits, was found to have the shape of a twisted heart with overall

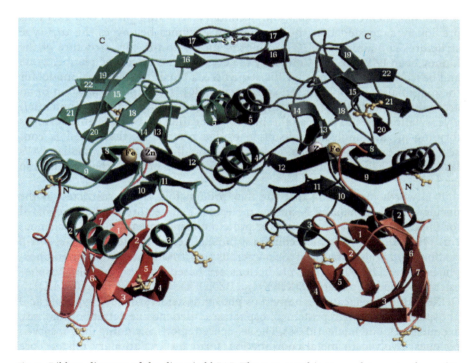

Fig. 2. Ribbon diagram of the dimeric kbPAP. The two metal ions are shown as spheres (Fe *yellow*, Zn *gray*) and the cysteine bridge (top, center) and the glycosylated asparagine side chains are drawn as ball and stick models (reprinted from Ref. [28]; copyright 1995 by the American Association for the Advancement of Science)

dimensions 40 by 60 by 75 Å (Fig. 2). The two dinuclear metal centers lie at the bottom of a broad pocket formed by both monomers and are exposed to the solvent.

Each monomer consists of two domains. The smaller N-terminal domain comprises about 120 amino acid residues (shown in red in Fig. 2) and forms a β-type structure. Surprisingly this domain does not participate in any interactions between the two subunits or with the active site. The alignment of the sequences from plant and mammalian PAPs suggests that this domain is absent in the monomeric 35 kDa mammalian enzymes [36].

The C-terminal α/β domain (residues 120 to 432, shown in green in Fig. 2) contains two sandwiched β-α-β-α-β structure motifs (1: $\beta8$-$\alpha1$-$\beta9$-$\alpha2$-$\beta10$; 2: $\beta12$-$\alpha4$-$\beta13$-$\alpha5$-$\beta14$) that form the core structure of the enzyme. The three parallel strands of both motifs are embedded in larger mixed β sheets consisting of 6 and 7 strands, respectively (1: $\beta10$-$\beta9$-$\beta8$-$\beta20$-$\beta21$-$\beta22$; 2: $\beta11$-$\beta12$-$\beta13$-$\beta14$-$\beta18$-$\beta15$-$\beta19$). The active site is located at the carboxy terminal site of the central β strands. All residues ligating the dimetal site are contributed from the loop regions following five of the six strands of the two central β-α-β-α-β motifs (see Fig. 2). According to the sequence alignment described above (shown in Fig. 1) all metal ligands are conserved between PAPs from mammalian, plant and bacterial sources.

Circular dichroism studies and secondary structure prediction methods previously suggested that the mammalian enzymes belong to the α/β-type structures containing two $\beta\alpha\beta\alpha\beta$ motifs similar to the core structure of the kidney bean PAP [36, 44]. Guided by the crystal structure of the plant enzyme and based on the alignment of the sequences, a tentative structural model for the mammalian PAPs has been constructed [36]. The exposed location of an accessible proteolytic cleavage site, the surface location of two (potential) glycosylation sites as well as the proximal location of two cysteine residues forming an intramolecular disulfide bridge reveal that the model is consistent with experimental data. These studies predict that the mammalian PAPs lack the four-helix-bundle motifs found in all di-iron proteins characterized so far.

It is worth mentioning that the new protein fold is not restricted to the purple acid phosphatases. A similar molecular architecture has been reported for the core structure of the Ser/Thr protein phosphatases 1 and 2A (calcineurin) [45–48], both of which contain a dimetal center located in a β-α-β-α-β scaffold, and is also predicted for numerous structurally uncharacterized phosphoesterases [49, 50]. These predictions are based on two "metallophosphoesterase" sequence patterns, DXH(X)$_{\sim25}$GDXXD(X)$_{\sim25}$GNH[E/D] and GH-(X)$_{\sim50}$-GHXH, which are located in numerous phosphoesterases, including purple acid phosphatases, Ser/Thr protein phosphatases, diadenosine tetraphosphatases, exonucleases, and nucleotidases. The sequence regions, including the conserved residues involved in metal coordination, form the evolutionary related core structure of these enzymes.

3
The Dimetal Site

3.1
Spectroscopic Characterization

3.1.1
UV/VIS and Resonance Raman Spectra

The dinuclear iron center of the mammalian PAPs exists in either of the two oxidation states: a reduced pink Fe(III)Fe(II) form and a purple oxidized Fe(III)Fe(III) form. The inactive diferric form exhibits a visible absorption maximum between 550 and 570 nm. The charge transfer band of the oxidized iron substituted form of the kidney bean enzyme peaks at approximately 560 nm [38], similar to the native Fe(III)Zn(II) form [20] and comparable to the mammalian diferric enzymes [1]. The absorption maximum of Fe(III)Fe(III) kbPAP, however, does not shift to 515 nm after reduction as observed for the mammalian PAPs.

First evidence about the nature of the unusual chromophore in PAPs has been provided by model studies on Fe(III)-phenolate complexes [51–53] and by comparison with other iron-proteins, such as transferrin [54] or protocatechuate 3,4-dioxygenase [55], suggesting that the intense visible absorption band is due to a phenolate-to-iron charge transfer transition. Because of the abnormally high extinction coefficient for one Fe(III)-tyrosine bond ($\varepsilon_M \sim 4000$, compared to values of 1000 to 2000 found for Fe(III)-model complexes), two or more tyrosyl ligands to the chromophoric iron have been suggested. However, no evidence of tyrosine heterogeneity has been found in the laser-Raman studies [56–58]. Excitation into the absorption band gives rise to four sharp resonance-enhanced tyrosine ring vibrations at 1164, 1287, 1498, and 1600 cm^{-1} (Fig. 3). The vibrational modes are observed for both, either purple or pink, Uf [56, 57] and bsPAP [58] and, therefore, reveal that the tyrosine coordinates with the iron ion that remains ferric in the reduced enzyme form (called chromophoric iron).

In addition to the tyrosinate modes discussed above, the Raman spectra of Uf and bsPAP also reveal four well-resolved low-frequency modes. Of particular interest is the band at 520 cm^{-1} (Uf) or 521 cm^{-1} (bsPAP), which shows a 2-fold increase upon reduction of the enzyme. Because of its sensitivity to the oxidation state of the protein but lack of ^{18}O dependence, that would be expected for an oxo or hydroxo bridge vibration, a possible assignment is the O-C-O bend of a bridging carboxylate coupled to an Fe-tyrosinate vibration. Furthermore, isotopic substitution experiments on pink and purple bsPAP failed to detect the vibrational mode of an Fe-O-Fe symmetric stretch, thus giving no direct evidence of an oxo-ligand.

3.1.2
EPR and Magnetic Measurements

Evidence for a dimetal center in the active sites of PAPs was initially provided by their EPR spectra. At temperatures below 30 K the mixed-valent di-iron

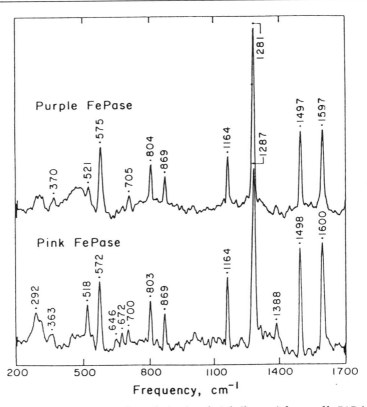

Fig. 3. Resonance Raman spectra of purple (*top*) and pink (*bottom*) forms of bsPAP (reprinted from Ref. [58]; copyright 1987 American Chemical Society)

forms of the native bovine spleen [58, 32] and porcine uterine enzymes [59, 60], as well as the zinc-exchanged di-iron kidney bean enzyme [37] exhibit a rhombic EPR signal with $g_{ave} \sim 1.7$ (Fig. 4). The low g_{ave} value is due to a $S = \Omega$ ground state characteristic of antiferromagnetically coupled Fe(III)Fe(II) sites initially found in hemerythrin [61], ribonucleotide reductase [62], and the two-iron ferredoxins [63]. The EPR signal has been found to be pH-dependent, resulting from a mixture of two species: a low pH form with g = 1.94, 1.78, and 1.65 and a high pH form with g = 1.85, 1.73, 1.58, related by an apparent pK_a of ~ 4.5 reported for the bovine spleen enzyme [58, 64]. Witzel et al. attributed this protonation equilibrium to a water molecule bound to the ferric iron (see section 4, Ref. [64]).

The native FeZn kidney bean enzyme has no EPR signal in the g = 1.7 range but shows a g = 4.3 signal characteristic of isolated high spin Fe(III) with large rhombicity [37]. However, the EPR spectrum of the Fe(III)Fe(II) derivative resembles those obtained for the mammalian PAPs showing that it contains an antiferromagnetically coupled dinuclear iron center. The completely restored catalytic activity for the Zn-exchanged di-iron form strongly suggests that the

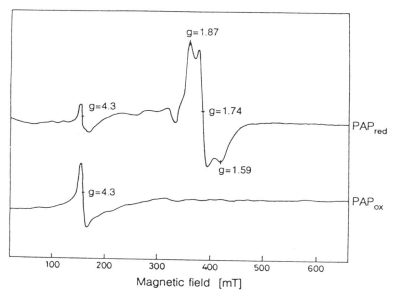

Fig. 4. EPR spectra of reduced (top) and oxidized (bottom) forms of bsPAP at pH 5.2 obtained at 5 K. Due to the antiferromagnetic coupling the oxidized Fe(III)Fe(III) form is EPR silent (S = 0). The small g = 4.3 signals account for a small percentage of ferric iron in the sample (reprinted from Ref. [64]; copyright 1991 Federation of European Biochemical Societies)

Zn(II) ion in the native plant enzyme occupies the same site as the Fe(II) ion in the di-iron derivative, thus supporting the existence of a dinuclear zinc-iron complex in kbPAP.

For the mammalian mixed-valent PAPs as well as for the Fe(III)Fe(II) derivative of kbPAP a small coupling constant (using the exchange Hamiltonian $H = JS_1S_2$) is revealed by EPR (11 cm^{-1} for bsPAP, Ref. [58]; 14 cm^{-1} for Uf, Ref. [65]), ^{1}H NMR (20 cm^{-1} for Uf, Ref. [66]), and SQUID (superconducting quantum interference device) measurements (19.8 cm^{-1} for Uf, Ref. [67]). Temperature-dependent magnetic susceptibility measurements on the Fe(III)Fe(II) derivative of kbPAP showed a similar exchange coupling of 15.9 cm^{-1} [68]. Studies on spin-coupled di-iron model compounds indicate that these values are consistent with the presence of a bridging μ-hydroxo or μ-phenoxo group [69]. For the EPR-silent oxidized diferric enzymes (S=0) a far greater antiferromagnetic interaction is reported (J = 300 cm^{-1} for bsPAP, Ref. [58]; J > 80 cm^{-1} for Uf, Ref. [66]), suggesting that oxidation might be accompanied by deprotonation of the μ-hydroxo bridge to a μ-oxo bridge. A more recent magnetic susceptibility study, however, reported a weak antiferromagnetic coupling (17.6 cm^{-1} for bsPAP and 18.0 cm^{-1} for kbPAP, Ref. [68]) indicating the lack of a μ-oxo bridge.

Replacement of the non-chromophoric ferrous iron in the mammalian enzymes with the diamagnetic Zn(II) ion eliminates the antiferromagnetic coupling. EPR signals at g = 4.3 are thus observed [70], similar to those found for the

native FeZn kidney bean enzyme [37], and typical of a high-spin Fe(III) center
(S = 5/2) with rhombic symmetry.

EPR measurements on the purple acid phosphatase from sweet potatoes led
to conflicting results regarding the metal content of these plant enzymes. The
EPR spectrum of the native enzyme from Kintoki potatoes has been reported to
be featureless, but upon denaturation a six-line signal around g = 2 appears
[71], characteristic of Mn(II). For the PAP from Jewel potatoes, however, a high-
spin Fe(III) signal similar to the spectrum of kbPAP has been observed suggesting that iron, rather than manganese, participates in catalysis and is the
visible chromophore [24].

3.1.3
ESEEM and ENDOR Spectra

ENDOR (electron nuclear double resonance) and ESSEM (electron spin echo
envelope modulation) spectroscopy are useful techniques for studying ligand
and substrate binding to the metal centers of paramagnetic metalloproteins.
Studies on uteroferrin reveal the presence of solvent-exchangeable protons
from an inner coordination sphere ligand. Some of these might be ascribed to
water molecules coordinated to the di-iron center [72, 73].

3.1.4
Mössbauer Spectra

^{57}Fe Mössbauer spectroscopy provides a direct probe of the electronic and
chemical environment of all of the iron atoms present in a sample. Mössbauer
data of ^{57}Fe-enriched Uf [59, 74, 75] and bsPAP [58, 76], as well as of Zn-exchanged ^{57}FeFe and ^{57}Fe^{57}Fe kbPAP [38] demonstrates that PAPs contain two
distinct, antiferromagnetically coupled high-spin iron atoms (S = $^1/_2$ for pink
enzyme, diamagnetic purple enzyme). The quadruple splittings of the ferric
sites are unusually large for a nominal S-state ion (Fig. 5), but not larger than
those found for the diferric centers of oxidized hemerythrin or ribonucleotide
reductase and related model compounds, and imply a strong distortion of the
octahedral symmetry imposed by the ligand structure of the dinuclear complex. The quadruple splitting seen for the chromophoric site in the pink enzyme
form is similar to that observed for the oxidized form, suggesting that the coordination geometry of this site is not influenced by the reduction of the enzyme.

3.1.5
EXAFS Spectra

Initial iron K-edge X-ray absorption measurements on the mammalian PAP
from bovine spleen revealed a metal-metal distance of 3.00 Å [77]. For the reduced form of uteroferrin Uf$_{red}$ analysis of the EXAFS data yielded two possible
Fe-Fe distances of 3.15 Å or 3.52 Å, compared to 3.21 Å for the oxidized form
complexed with phosphate [78]. Given the absence of a 1.8 Å Fe-bond, the ferric

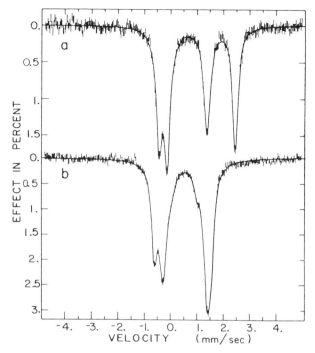

Fig. 5. Mössbauer spectra of the ^{57}Fe-enriched pink form of uteroferrin at 185 K (A) and of the purple form at 10 K (B). The *solid lines* are Lorentzian fits obtained with the parameters listed in Ref. [59] (reprinted from Ref. [59]; copyright 1983 Elsevier Science Publishers B.V.)

site of these enzymes is not oxo-bridged in agreement with the resonance Raman results described before. The near-edge absorption data suggest that both iron atoms in purple and pink bovine spleen PAP are coordinated to N/O-donors in six-coordinate sites with relatively low symmetry.

The metal-metal distance of 3.9 Å determined for kbPAP by EXAFS measurements [79] is not only significantly larger than the value for the mammalian enzymes but also inconsistent with the metal separation revealed by crystallographic studies (see Sect. 3.2). More recent EXAFS studies, however, indicate that the metal-metal distance in the Fe(III)Zn(II) form resembles that of Uf and bsPAP and that the experimental conditions chosen for previous EXAFS measurements led to the photoreduction of the chromophoric iron [80].

3.1.6
^{1}H-NMR Spectra

^{1}H-NMR spectroscopy has proven to be a powerful tool for probing the metal environment of many paramagnetic metalloproteins by analysing the isotropically shifted resonances arising from the nuclei in close proximity to the metal

[81]. Several residues ligating the dimetal center of PAPs have been identified by proton NMR studies on uteroferrin [66, 82 – 84], bovine spleen enzyme [83], and Zn-exchanged Fe(III)Fe(II)kbPAP [85]. These studies demonstrate that despite the small differences in the local coordination environments, the types of ligands, coordination number, and coordination geometry associated within the dimetal active site of plant and mammalian PAPs are the same. Most of the observed ^{1}H-NMR resonances (shown in Fig. 6 for Uf) could be assigned on the basis of their chemical shifts, the presence or absence of the peaks in D_2O, distance estimates based on the T_1 values, and the observation of cross relaxation (nuclear Overhauser effect).

Signals a, c, d, j, and y are all associated with a tyrosine bound to Fe(III), as previously demonstrated by resonance Raman studies. Whereas the ortho protons exhibit a large upfield shift (signal y, not shown in Fig. 6), the meta protons

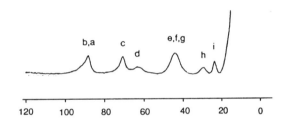

Fig. 6. Proton NMR spectrum (downfield range) of reduced Uf_{red} at 300 MHz and 30 °C in 100 mM acetate buffer at pH 4.9 and resulting model for the dinuclear iron center of mammalian PAPs. Ligands with *dashed bonds* have not been unequivocally established (both figures are adapted from Ref. [83])

show resonances between 62 and 70 ppm in the downfield range (c and d). The presence of two peaks for the meta protons in the spectra for Uf is ascribed to a slow rotation of the tyrosine about the C_β-C_γ bond, approaching coalescence at 50 °C. For the bovine spleen and the kidney bean enzyme these signals are not resolved, suggesting a significantly lower coalescence temperature. In uteroferrin, the tyrosine β-CH_2 protons resonate at $\delta = 87$ ppm (signal a) and 15 ppm (signal j), compared to $\delta = 88$ ppm and 13 ppm for kbPAP and $\delta = 70$ ppm for one of the C_β protons of bsPAP. This suggests that the conformation of the Fe(III)-coordinated tyrosine residue is very similar in Uf and kbPAP, but not identical to that of bsPAP.

Comparison of the NMR spectra of Uf in H_2O and D_2O reveals the presence of three solvent-exchangeable resonances at 87 ppm (signal b), 44 ppm (signal e), and -25 ppm (signal ×) in the upfield range (not shown in Fig. 6). The first two features are also present in the spectra from bsPAP and di-iron kbPAP and, on the basis of comparisons with model complexes, have been assigned to the NH protons of a N_ε-coordinated histidine coordinated to Fe(III) and a N_δ-coordinated histidine coordinated to iron(II), respectively. Significant NOE correlation between the signals b and i and between the signals e and g supports the assignment of the features i and g as the C_β proton of the Fe(III)-coordinated histidine residue and the C_δ-proton of the Fe(II)-coordinated histidine residue, respectively (see Fig. 6). The upfield signal × seen in Uf (due to the broadness not observed in the spectrum of bsPAP) was arbitrarily attributed to an amide proton in a complex where the amide carbonyl oxygen is coordinated to the metal.

The remaining signals, f and h, appear to arise from the carboxymethylene protons of a bridging carboxylate ligand and an aspartate/glutamate residue coordinated to Fe(II). ^1H NOESY studies on Fe(II)-exchanged Fe(III)Co(II)Uf revealed the presence of an ABX spin system assigned to a YCH_2CH amino acid side-chain with Y ligated to the metal center [84]. Based on the isotropic shifts and the T_1 values for the corresponding signals, the pattern was attributed to an aspartate/glutamate bound to the Co(II) center of Fe(III)Co(II)Uf. However, these findings are equally consistent with a Me(II)-coordinated asparagine residue Asn-201, found in the X-ray crystal structure of the FeZn kidney bean enzyme. Asparagine ligation via the carbonyl group to the ferrous iron in native mammalian PAPs also allows the attribution of the upfield signal × to the nearby amide proton (see above).

3.2
The Crystal Structure of the Fe(III)Zn(II) Unit of kbPAP

The crystal structure of the kidney bean enzyme provides a complete 3-dimensional picture of the dimetal center of this enzyme class (see Fig. 7) [28, 29]. The ferric iron is coordinated by Tyr-167, by the N_ε atom of His-325, and monodentately by the carboxylate group of Asp-135. The zinc ion is ligated by the N_ε atom of His-286, the N_δ atom of His-323 and the amide oxygen of Asn-201.

The two metal ions are bridged monodentately by the carboxylate group of Asp-164. At a resolution of 2.65 Å, solvent molecules coordinated to the dimetal

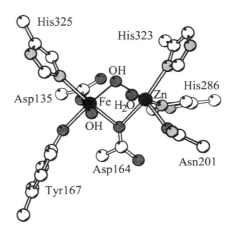

Fig. 7. Dinuclear metal center of Fe(III)Zn(II) kidney bean PAP displaying the metal coordinating ligands. The three solvent ligands have been placed solely on the basis of the observed coordination geometry around the metal ions. Their presence is based on spectroscopic and kinetic studies (figure adapted from Ref. [28])

center could not be clearly located in the electron density maps. On the basis of the spectroscopic data reported above (see section 3.1) and in agreement with the proposed general reaction path (see section 4), three additional ligands were included in the active site structure. A hydroxide ion ligated to Fe(III), a bridging hydroxide ion, and a water molecule coordinated to Zn(II) complete the coordination spheres of both metals, giving an octahedron for Fe(III) and a distorted octahedron for Zn(II). The distance between the two metal ions has been refined to 3.26 Å [29].

The paramagnetically shifted protons in the residues corresponding to His-323, His-325, and Tyr-167 have been identified in the spectra from the mammalian enzymes and the signals caused by the asparagine coordinated to the divalent metal can be assigned. However, the ^{1}H-NMR spectra of the mammalian PAPs [66, 82–84] and the proton NMR spectrum of di-iron kbPAP [85] show no evidence for the coordination of a second histidine residue to the ferrous iron (His-286 in kbPAP). A fast exchange of the N_ε proton with the bulk solvent has been considered as an explanation for the absence of its signal in the proton NMR spectrum. However, its limited solvent accessibility does not support this assumption.

3.3
Interaction with Phosphate and Other Tetrahedral Oxoanions

Purple acid phosphatases exhibit product inhibition by phosphate ($K_i \sim$ 3.4–5.4 mM) as well as inhibition by other tetrahedral oxoanions, including arsenate ($K_i \sim 0.18$–0.64 mM), vanadate ($K_i \sim 4$ µM), molybdate ($K_i \sim 1.9$–5.0 µM) and tungstate ($K_i \sim 0.7$–10 µM) [44, 86]. Studies on the complexes of the mammalian enzymes with phosphate were initially complicated by both the peculiar spectroscopic properties and the increased susceptibility of the Fe(III)Fe(II) site to oxidation upon binding phosphate. After several conflicting reports, the formation of an Fe(III)Fe(II) · PO_4 intermediate could finally be demonstrated by Mössbauer [74] and EPR spectroscopy [67]. These studies support rapid binding of phosphate to the enzyme, followed by slow oxidation and loss of activity.

Monodentate or bidentate bridging binding mode of phosphate? The binding of phosphate to the mixed-valent iron center reduces the coupling constant (J = 6.0 cm^{-1} determined for Uf by SQUID measurements [67]) causing a broad EPR signal which is difficult to detect under the conditions used for the phosphate-free reduced enzyme form. The weakening of the antiferromagnetic coupling in Uf$_{red}$ upon binding phosphate is proposed to result from an interaction of the phosphate ion with the hydroxo bridge. Effects of H_2O/D_2O substitution on the EPR spectrum of Uf$_{red}$ · AO$_4$ support a model wherein the binding of phosphate or arsenate entails at least some proton transfer from the anion to the hydroxo bridge [70]. EPR experiments with ^{17}O-labeled phosphate on Fe(III)Zn(II)Uf showed broadening of the g = 4.3 resonance, characteristic of a hyperfine interaction of the ferric iron with ^{17}O (I = 5/2). The observation of hyperfine broadening provided direct evidence for phosphate as a ligand for the Fe(III) site in Fe(III)Zn(II) · PO_4.

Binding of phosphate to the chromophoric ferric iron, but not to the ferrous iron, was also concluded from Mössbauer data [74], which show significantly greater changes of the ferric, rather than the ferrous doublet on binding phosphate. Mössbauer measurements on the Zn-exchanged di-iron kbPAP, however, show only minor differences for both iron ions in the reduced mixed valent and the oxidized form upon addition of phosphate [38]. The conclusion that in the reduced form phosphate binds only to the ferric ion on Uf, as indicated by Mössbauer data, is thus not supported by data obtained for kbPAP.

Some authors have suggested a stronger bidentate binding at acid pH and a weaker monodentate binding at basic pH, on the basis of the pH dependence of the inhibition constant for phosphate, K_i, as well as the pH-dependent shift of λ_{max} of the visible spectra after phosphate addition [74] (both conversions show an apparent pK_a value of around 4.9) [64]. However, the Mössbauer spectrum at pH 4.9 does not provide any evidence for two forms of the reduced phosphate complex [74]. Furthermore, the EPR spectrum of the Fe(III)Zn(II)kbPAP phosphate complex (which shows a higher symmetry in the coordination sphere of the ferric iron upon phosphate binding) reveals no pH dependence.

At present there is no direct experimental evidence that can exclude the binding of phosphate in a bidentate bridging mode to the mixed-valent dimetal unit, and given the number of synthetic di-iron complexes with phosphate bridges a bridging phosphate is certainly conceivable in the case of Uf [87–90]. However, if phosphate or arsenate do bridge the di-iron center, how would molybdate or tungstate coordination (see below) differ to cause their much greater binding affinity?

Electrochemical studies showed that phosphate binding makes the reduction potential of Uf more negative by 193 mV [91] and renders the enzyme more susceptible to aerobic oxidation. For the resulting oxidized Fe(III)Fe(III) · PO$_4$ form, a bidentate bridging mode has been revealed by EXAFS studies on utero-ferrin. In presence of phosphate, the Fe-Fe distance was found to be 3.21 Å [78]. The observed Fe-P (3.17 Å) distance corresponds to Fe-O-P angles of 129° which are indicative of a bidentate bridging oxoanion.

In contrast to phosphate, molybdate and tungstate transform the rhombic EPR spectrum of Uf to a sharp axial signal [73, 74, 92]. Furthermore, molybdate makes the potential of Uf more positive by 192 mV, thus stabilizing the mixed-valent center to air oxidation [91]. The subsequent addition of phosphate to the Fe(III)Fe(II) · MO$_4$ complex has no effect on the EPR spectra, strongly suggesting that molybdate blocks the phosphate binding site and both tetrahedral oxoanions compete for the same binding site on the protein [72, 73].

The bidentate bridging binding mode of both phosphate and tungstate to the heterodinuclear Fe-Zn metal center of kbPAP has been recently revealed by the crystal structures of this enzyme complexed with these inhibitors, at a resolution of 2.7 and 3.0 Å, respectively [29]. In both cases, anion binding is not accompanied by any large conformational change in the enzyme structure. Small movements with a maximal coordinate shift of 1 Å are only observed for two of the active site residues viz. His-295 and His-296. Phosphate, as well as tungstate, replace two of the presumed solvent ligands present in the inhibitor-free enzyme form (Fig. 8). On binding to the inhibitors the metal-metal distance changes marginally to 3.33 Å in the complex with phosphate and to 3.20 Å in

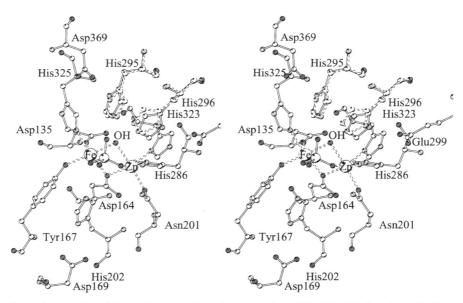

Fig. 8. Stereo view of the active site region of the complex of kbPAP with phosphate displaying the bidentate binding mode of the phosphate ion. Carbon atoms are colored *white*, oxygen atoms and nitrogen atoms *gray*. The two histidine residues His-295 and His-296 in their conformation found for the phosphate-free enzyme are depicted in *broken lines* (reprinted from Ref. [29]; copyright 1995 Academic Press)

the enzyme complex with tungstate. The bridging mode of both oxoanions is nearly symmetrical with Fe-O-P(W) and Zn-O-P(W) angles of $120°$ ($112°$) and $127°$ ($117°$), respectively. The 2.8 Å iron-phosphorus and 3.1 Å zinc-phosphorus distances reported in the enzyme complex with phosphate are remarkably short. However, because of the estimated coordinate error of 0.25 Å they are still consistent with the 3.17 Å value found by EXAFS measurements on Fe(III)Fe(III)Uf · PO_4 [78]. Unfortunately, the limited resolution of the present crystallographic studies does not allow the identification of minor structural differences in the binding mode of these oxoanions, which could explain the differences in their inhibition constants and the spectroscopic results, including EPR, Mössbauer and UV/Vis.

4
The Proposed Reaction Path: Two-Metal Ion Catalysis by PAPs

Phosphatases may be classified with respect to their reaction, depending upon whether the phosphoryl group of the substrate is directly transferred to a water molecule or a nucleophilic amino acid residue of the enzyme attacks the phosphorus to form a phosphoenzyme intermediate that is subsequently hydrolyzed. The phosphoenzyme pathway implies the existence of an intermediate, which may be trapped at a certain pH. Initial studies on PAPs provided indirect evidence for the formation of a covalent phosphoryl-enzyme intermediate [93]. However, more recent studies fail to isolate an intermediate [94].

The stereoochemical course of PAP catalysis was probed by analyzing the products of the hydrolysis of the chiral substrate S_P-$2',3'$-methoxymethylidene-$ATP_γS_γ{}^{18}O_γ{}^{17}O$ by ^{31}P NMR [95]. The observed overall inversion of the configuration at the phosphorus atom rules out a phosphoenzyme pathway and supports the direct transfer of the phospho group to water in an in-line S_N2-type displacement step. Kinetic data show a typical bell-shaped dependence of V_{max} on pH with an optimum of pH 5.9 and two apparent pK_a values of 4.8 and 6.9 using $α$-naphthyl phosphate as substrate [64]. The lower pK_a has been attributed to the protonation equilibrium of an Fe(III)-coordinated solvent molecule assuming that it forms the nucleophilic hydroxide ion that attacks the phosphorus of the substrate [64]. This attribution is also supported by pH-dependent EPR measurements which reveal a well matching pK_a value of 4.4 (see section 3.1.2). Evidently, the activation of the water molecule by coordination to the ferric iron allows the formation of more nucleophilic hydroxide ions even at low pH. For comparison, the first dissociation constant for $[Fe(H_2O)_6]^{3+}$ is 3.0 [96].

What role does the divalent metal ion play in the phosphatase mechanism? Stopped-flow measurements on uteroferrin indicate that a rapid binding of the substrate to Fe(II) precedes the rate-controlling attack of the nucleophilic hydroxide ion [97, 98]. In a plausible mechanism that is based on the structures of kbPAP and its complex with phosphate (as shown in Fig. 9) the substrate is bound by Zn(II) (corresponding to Fe(II) in the mammalian enzymes) and two histidine residues His-202 and His-296, both of which are conserved in the mammalian PAPs (corresponding to His-92 and His-193 in Uf, Ref. [36]). In addition, the amide group of the asparagine residue coordinated to the divalent

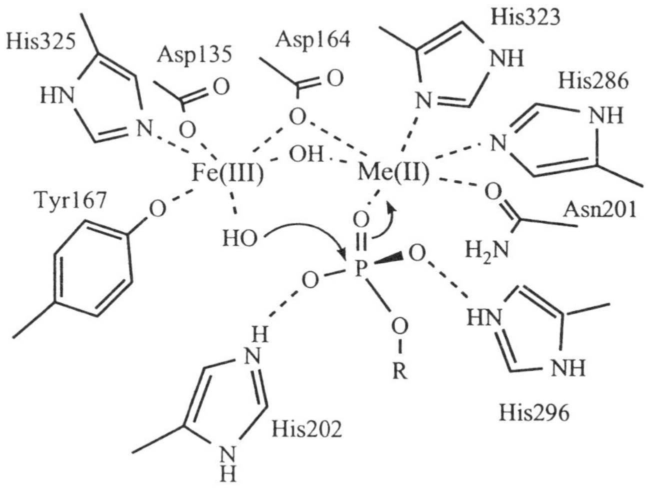

Fig. 9. Tentative mechanism for the hydrolysis of phosphomonoesters by PAP (figure adapted from Ref. [29])

metal ion might form a hydrogen bond with the Me(II)-bound phosphoryl oxygen atom. Coordination of an aspartate residue, a more commonly observed metal ligand, at this position would most likely result in repulsive forces between the negatively charged carboxylate group and the phosphoryl group of the substrate. The present interactions are supposed to preorient the phosphate group of the substrate for an in-line attack of the Fe(III)-bound hydroxide ion from a direction opposite the esterified oxygen. The configuration at the phosphorus is inverted via a pentacoordinate transition state in which the leaving group and the attacking nucleophile are in apical positions. In addition to the metal ions, the protonated side chains of His-202 and His-296 promote catalysis by reducing the energy of the pentacovalent transition state. The decrease of V_{max} at alkaline pH with an apparent pK_a at 6.9 can be attributed to the deprotonation of one of these histidine residues resulting in an increased transition state energy and a reduced reaction rate. Subsequently the alcohol group of the substrate is released generating a phosphate ion bound to both metal ions via two oxygen atoms.

In future, site-directed mutagenesis might probe the function of individual active site residues involved in the mechanism that has been proposed on the basis of spectroscopic and crystallographic studies as reviewed in this article. The recently established baculovirus vector expression systems for the human type 5 acid phosphatase [99] and for rat bone PAP [100] will allow the engineering of these enzyme mutants. Furthermore, a structure determination of any plant or mammalian PAP at higher resolution is desirable in order to determine the exact position of solvent molecules coordinated to the dimetal site, that are supposed to play a cardinal role in the hydrolysis reaction catalyzed by this class of enzyme.

5
The Enigma: The Physiological Role of PAP/TRAPs

Despite extensive knowledge about the substrate specificity, inhibitor sensitivity, subcellular localization, active site structure, relationship to other enzymes, DNA and protein sequence, and the chemical mechanism, the physiological function of PAP/TRAPs continues to be obscure. The mammalian PAPs may, in fact, have multiple functions, with different properties in different cell types. The purple acid phosphatase from spleen has been proposed to play a role in the degradation of aged erythrocytes [101]. The function might be correlated to the protein phosphatase activity of the enzyme degrading phosphoproteins present on the erythrocyte membrane. The detection of peroxidase activity for human PAP in vitro [99] has led to the suggestion that the degradative function is correlated to the generation of oxygen-derived free radicals by the divalent iron center. At the same time, the physiological function of secreted mammalian bone PAP in osteoclasts is much better understood. There is experimental evidence for the involvement of this enzyme in the dephosphorylation of the bone matrix phosphoproteins viz. osteopontin and bone sialoprotein and thus in the regulation of osteoclast attachment to bone [102]. Unlike the protein phosphatases with regulatory functions the substrate specificity of osteoclast PAP is unusually broad. Control of the enzyme activity in such a case, therefore, could be achieved by factors such as local pH, redox conditions, the valency of the dinuclear metal center and the local phosphate concentration.

For the plant PAPs a physiological function in the liberation of phosphate from organophosphates, especially after phosphate starvation, has been recently proposed for the enzyme secreted from *Arabidopsis thaliana*. Increased levels of mRNA for this protein in response to phosphate-starvation support this hypothesis [27]. Efforts to prove a function of the plant PAPs as unspecific phosphatases using transgenic plants are in progress.

Acknowledgments. We are grateful to Gianantonio Battistuzzi and Thomas D. McKnight for sharing results before publication and Gianantonio Battistuzzi for his helpful comments. We would also like to thank James C. Sacchettini and his lab for linguistic assistance. Research in the laboratory of B.K. was supported by grants DFG 406/13/15 and BMFT(BMBF)05648PMA. T.K. thanks the Deutsche Forschungsgemeinschaft for financial support.

6
References

1. Campbell HD, Zerner B (1973) Biochem Biophys Res Commun 54:1498
2. Chen TT, Baser FW, Cetorelli JJ, Pollard WE, Roberts RM (1973) J Biol Chem 248:8560
3. Doi K, Antanaitis BC, Aisen P (1988) Struct Bonding 70:1
4. Que Jr L, True AE (1990) Prog Inorg Chem 38:97
5. Vincent JB, Olivier-Lilley GL, Averill BA (1990) Chem Rev 90:1447
6. Li CY, Yam LT, Lam KW (1970) J Histochem Cytochem 18:473
7. Li CY, Yam LT, Lam KW (1970) J Histochem Cytochem 18:901
8. Yam LT, Li CY, Lam KW (1971) N Engl J Med 284:357
9. Ketcham CM, Roberts RM, Simmen RCM, Nick HS (1989) J Biol Chem 264:557
10. Ketcham CM, Baumbach GA, Bazer FW, Roberts RM (1985) J Biol Chem 260:5768

11. Allen BS, Nuttleman PR, Ketcham CM, Roberts RM (1989) J Bone Miner Res 4:47
12. Efstratiadis T, Moss DW (1985) Enzyme 33:34
13. Schindelmeiser J, Schewe P, Zonka T, Münstermann D (1989) Histochemistry 92:81
14. Lau K-HW, Freeman T, Baylink DJ (1987) J Biol Chem 262:1389
15. Hara A, Sawada H, Kato T, Nakayama T, Yamamoto H, Matsumoto Y (1984) J Biochem 95:67
16. Andersson GN, Ek-Rylander B, Hammarström LE (1984) Arch Biochem Biophys 228:431
17. Hara A, Kato T, Sawada H, Fukuyama K, Epstein WL (1985) Comp Biochem Physiol 82B:269
18. Nordlund P, Eklund H (1995) Curr Opin Struct Biol 5:758
19. Nochumson S, O'Rangers JJ, Dimitrov NV (1974) Fed Proc 33:1378
20. Beck JL, McConachie LA, Summors AC, Arnold WN, de Jersey J, Zerner B (1986) Biochim Biophys Acta 869:61
21. Fujimoto S, Nakagawa T, Ohara A (1977) Agric Biol Chem 41:599
22. LeBansky BR, McKnight TD, Griffing LR (1992) Plant Physiol 99:391
23. Uehara K, Fujimoto S, Taniguchi T (1974) J Biochem 75:627
24. Hefler SK, Averill BA (1987) Biochem Biophys Res Commun 146:1173
25. Fujimoto S, Nakagawa T, Ishimitsu S, Ohara A (1977) Chem Pharm Bull 25:1459
26. Igaue I, Watabe H, Takahashi K, Takekoshi M, Morota A (1976) Agric Biol Chem 40:823
27. Patel KS, Lockless SW, McKnight TD (1997) Plant Mol Biol (in press)
28. Sträter N, Klabunde T, Tucker P, Witzel H, Krebs B (1995) Science 268:1489
29. Klabunde T, Sträter N, Fröhlich R, Witzel H, Krebs B (1996) J Mol Biol 259:737
30. Beck JL, McArthur MJ, de Jersey J, Zerner B (1988) Inorg Chim Acta 153:39
31. Beck JL, Keough DT, de Jersey J, Zerner B (1984) Biochim Biophys Acta 791:357
32. Davis JC, Averill BA (1982) Proc Natl Acad Sci USA 79:4623
33. Ullah AHJ, Mullaney EM, Dischinger HC Jr (1994) Biochem Biophys Res Commun 203:182
34. Jacobs MM, Nyc JF, Brown DM (1971) J Biol Chem 246:1419
35. Glew RH, Heath EC (1971) J Biol Chem 246:1556
36. Klabunde T, Sträter N, Krebs B, Witzel H (1995) FEBS Lett 367:56
37. Beck JL, de Jersey J, Zerner B, Hendrich MP, Debrunner PG (1988) J Am Chem Soc 110:3317
38. Suerbaum H, Körner M, Witzel H, Althaus E, Mosel B-D, Müller-Warmuth W (1993) Eur J Biochem 214:313
39. Hunt DF, Yates JR III, Shabanowitz J, Zhu N-Z, Zirino T, Averill BA, Daurat-Larroque ST, Shewale JG, Roberts RM, Brew K (1987) Biochem Biophys Res Commun 144:1154
40. Lord DK, Cross NCP, Bevilacqua MA, Rider SH, Gorman PA, Groves AV, Moss DW, Sheer D, Cox TM (1990) Eur J Biochem 189:287
41. Ek-Rylander B, Bill P, Norgård M, Nilsson S, Andersson G (1991) J Biol Chem 266:24684
42. Cassady AI, King AG, Cross NCP, Hume DA (1993) Gene 130:201
43. Klabunde T, Stahl B, Suerbaum H, Hahner S, Karas M, Hillenkamp F, Krebs B, Witzel H (1994) Eur J Biochem 226:369
44. Vincent JB, Crowder MW, Averill BA (1991) Biochemistry 30:3025
45. Goldberg J, Huang H, Kwon Y, Greengard P, Nairn AC, Kuriyan J (1995) Nature 376:745
46. Griffith JP, Kim JL, Kim EE, Sintchak MD, Thomson JA, Fitzgibbon MJ, Fleming MA, Caron PR, Hsiao K, Navia MA (1995) Cell 82:507
47. Kissinger CR, Parge HE, Knighton DR, Lewis CT, Pelletier LA, Tempczyk A, Kalish VJ, Tucker KD, Showalter RE, Moomaw EW, Gastinel LN, Habuka N, Chen X, Maldonado F, Barker JE, Bacquet R, Villafranca JE (1995) Nature 378:641
48. Egloff M, Cohen PT, Reinemer P, Barford D (1995) J Mol Biol 254:942
49. Koonin EV (1994) Protein Sci 3:356
50. Zhou S, Clemens JC, Stone RL, Dixon JE (1994) J Biol Chem 269:26234
51. Ackerman GA, Hesse D (1970) Z Anorg Allg Chem 375:77
52. Gaber BP, Miskowski V, Spiro TG (1974) J Am Chem Soc 96:6868
53. Ainscough EW, Brodie AM, Plowman JE, Brown KL, Addison AW, Gainsford AR (1980) Inorg Chem 19:3655

54. Wood JM, Lipscomb JD, Que Jr L., Stephens RS, Orme-Johnson WH, Münck E, Ridley WP, Dizikes L, Cheh A, Francia M, Frick T, Zimmerman R, Howard J (1977) Some bio-inorganic chemical reactions of environmental significance. In: Addison AW, Cullen WR, Dolphin D, James BR (eds) Biological Aspects of Inorganic Chemistry. Wiley-Interscience, New York. p 261
55. Tatsuno Y, Saecki Y, Iwaki M, Yagi T, Nozaki M, Kitagawa T, Otsuka S (1978) J Am Chem Soc 100:4614
56. Gaber BP, Sheridan JP, Bazer FW, Roberts RM (1979) J Biol Chem 254:8340
57. Antanaitis BC, Strekas T, Aisen P (1982) J Biol Chem 257:3766
58. Averill BA, Davis JC, Burman S, Zirino T, Sanders-Loehr J, Loehr TM, Sage JT, Debrunner PG (1987) J Am Chem Soc 109:3760
59. Debrunner PG, Hendrich MP, de Jersey J, Keough DT, Sage JT, Zerner B (1983) Biochim Biophys Acta 745:103
60. Antanaitis BC, Aisen P, Lilienthal HR, Roberts RM, Bazer FW (1980) J Biol Chem 255:11204
61. Muhoberac BB, Wharton DC, Babcock LM, Harrington PC, Wilkins RG (1980) Biochim Biophys Acta 626:37
62. Petersson L, Graslund A, Ehrenberg A, Sjöberg BM, Reichard P (1980) J Biol Chem 255:6706
63. Sands RH, Dunham WR (1975) Quart Rev Biophys 7:443
64. Dietrich M, Münstermann D, Suerbaum H, Witzel H (1991) Eur J Biochem 199:105
65. Antanaitis BC, Aisen P, Lilienthal HR (1983) J Biol Chem 258:3166
66. Lauffer RB, Antanaitis BC, Aisen P, Que Jr L (1983) J Biol Chem 258:14212
67. Day EP, David SS, Peterson J, Dunham WR, Bonvoisin JJ, Sands RH, Que Jr L (1988) J Biol Chem 263:15561
68. Gehring S, Fleischhauer P, Behlendorf M, Hübner M, Lorösch J, Haase W, Dietrich M, Löcke R, Krebs B, Witzel H (1996) Inorg Chim Acta 252:13
69. Kurtz D Jr (1990) Chem Rev 90, 585–606
70. David SS, Que Jr L (1990) J Am Chem Soc 112:6455
71. Sugiura Y, Kawabe H, Tanaka H, Fujimoto S, Ohara A (1981) J Biol Chem 256:10664
72. Doi K, McCracken J, Peisach J, Aisen P (1988) J Biol Chem 263:5757
73. Antanaitis BC, Aisen P (1985) J Biol Chem 260:751
74. Pyrz JW, Sage JT, Debrunner PG, Que Jr L (1986) J Biol Chem 261:11015
75. Sage JT, Xia Y-M, Debrunner PG, Keough, DT, de Jersey J, Zerner B (1989) J Am Chem Soc 111:7239
76. Cichutek K, Witzel H, Parak F (1988) Hyperfine Interactions 42:885
77. Kauzlarich SM, Teo BK, Zirino T, Burman S, Davis JC, Averill BA (1986) Inorg Chem 25:2781
78. True AE, Scarrow RC, Randall CR, Holz RC, Que Jr L (1993) J Am Chem Soc 115:4246
79. Priggemeyer S, Eggers-Borkenstein P, Ahlers F, Henkel G, Körner M, Witzel H, Nolting H-F, Hermes C, Krebs B (1995) Inorg Chem 34:1445
80. Krebs B, Sift B (personal communication)
81. Bertini I, Luchinat C (1986) NMR of Paramagnetic Molecules in Biological Systems, Benjamin/Cummings, Menlo Park
82. Scarrow RC, Pyrz JW, Que Jr L (1990) J Am Chem Soc 112:657
83. Wang Z, Ming L-J, Que Jr L (1992) Biochemistry 31:5263
84. Holz RC, Que Jr L, Ming L-J (1992) J Am Chem Soc 114:4434
85. Battistuzzi G, Dietrich M, Löcke R, Witzel H (1997) Biochem J 323:593
86. Crans DC, Simone CM, Holz RC, Que Jr L (1992) Biochemistry 31:11731
87. Schepers K, Bremer B, Krebs B, Henkel G, Althaus E, Mosel B, Müller-Warmuth W (1990) Angew Chem 102:582; Angew Chem Int Ed Engl 29:531
88. Krebs B, Schepers K, Bremer B, Henkel G, Althaus E, Müller-Warmuth W, Griesar K, Haase W (1994) Inorg Chem 33:1907
89. Eulering B, Ahlers F, Zippel F, Schmidt M, Nolting H-F, Krebs B (1995) J Chem Soc Chem Comm:1305

90. Jang HG, Hendrich MP, Que Jr L (1993) Biochemistry 32:911
91. Wang DL, Holz RC, David SS, Que Jr L, Stankovich MT (1991) Biochemistry 30:8187
92. Crowder MW, Vincent JB, Averill BA (1992) Biochemistry 31:9603
93. Vincent JB, Crowder MW, Averill BA (1991) J Biol Chem 266:17737
94. Wynne CJ, Hamilton SE, Dionysius DA, Beck JL, de Jersey J (1995) Arch Biochem Biophys 319:133
95. Mueller EG, Crowder MW, Averill BA, Knowles JR (1993) J Am Chem Soc 115:2974
96. Baes CF, Mesmer E (1976) The Hydrolysis of Cations, Wiley-Interscience, New York
97. Aquino MAS, Lim J-S, Sykes AG (1994) J Chem Soc Dalton Trans:429
98. Aquino MAS, Lim J-S, Sykes AG (1992) J Chem Soc Dalton Trans:2135
99. Hayman AR, Cox TM (1994) J Biol Chem 269:1294
100. Ek-Rylander B, Berkhem T, Husman B, Oehman L, Andersson KK, Andersson G (unpublished results)
101. Schindelmeiser J, Münstermann D, Witzel H (1987) Histochemistry 87:13
102. Ek-Rylander B, Flores M, Wendel M, Heinegård D, Andersson, G (1994) J Biol Chem 269:14853

Springer
and the
environment

 Springer

Printing: Saladruck, Berlin
Binding: Buchbinderei Lüderitz & Bauer, Berlin